12. COLLOQUIUM DER
GESELLSCHAFT FÜR PHYSIOLOGISCHE CHEMIE
AM 13./15. APRIL 1961 IN MOSBACH/BADEN

BIOCHEMIE DES AKTIVEN TRANSPORTS

MIT 69 TEXTABBILDUNGEN

Springer-Verlag Berlin Heidelberg GmbH

ISBN 978-3-662-01363-2 ISBN 978-3-662-01362-5 (eBook)
DOI 10.1007/978-3-662-01362-5

Ursprünglich erschienen bei Springer-Verlag OHG. Berlin · Göttingen · Heidelberg 1961.

Inhalt

Begrüßung und Eröffnung

Klenk: Meine Damen und Herren. Es ist mir eine Freude, Sie alle zum 12. Mosbacher Colloquium hier zu begrüßen, besonders auch unsere ausländischen Gäste: die Herren Ussing, Wilbrandt, Keynes und Kleinzeller. Der Vortrag von Herrn Mitchell muß leider ausfallen, da der Redner erkrankt ist. An seiner Stelle spricht aber Herr Professor Kepès aus dem Pasteur Institut in Paris. Wir sind ihm ganz besonders dankbar dafür, daß er sich bereit erklärt hat, so kurzfristig einzuspringen.

Die Organisation des Symposiums lag in den Händen der Herren Netter und Heinz. Auch Ihnen gilt unser besonderer Dank.

In Mosbach war es von jeher üblich, frei und offen und frisch von der Leber weg zu diskutieren. Wir haben uns bisher immer damit begnügt, die Diskussion einfach auf ein Tonband aufzunehmen. Leider hat es sich herausgestellt, daß die redaktionelle Bearbeitung einer solchen Tonbandaufnahme doch erhebliche Schwierigkeiten macht, und so möchte ich die Diskussionsredner bitten, zu unserer Erleichterung nun ihre Bemerkungen zusätzlich noch schriftlich zu fixieren und nach der Diskussion dann den Zettel hier abzugeben.

Das Wort erhielt anschließend der Herr Bürgermeister der Stadt Mosbach zu einer Begrüßungsansprache. — Außerdem gedachte Herr Prof. Ackermann *noch des Todes von Professor* Felix, *des langjährigen Vorsitzenden der physiologisch-chemischen Gesellschaft, mit folgenden Worten:*

Ackermann (Würzburg): Liebe Freunde, ich glaube wir müssen auch noch einmal des lieben alten Freundes Kurt Felix gedenken. Es ist dies zwar schon in Zürich erfolgt, aber Kurt Felix ist eng verbunden mit diesem kleinen erfolgreichen Kreis. Er ist ja der Gründer davon, so daß ich der Meinung bin, wir müssen noch einmal seiner gedenken. Und ich habe mir die Freiheit genommen, dies zum Ausdruck zu bringen, weil Kurt Felix auch noch mein Schüler war. Allerdings nicht Schüler in dem eigentlichen Sinne, daß ich Einfluß auf seine Arbeiten gehabt hätte. Er war jedoch bei mir im Laboratorium in Würzburg als Student und stellte Tyrosin aus Hornspänen dar. Es war also schon eine gewisse Tendenz zum Eiweiß. Und schon damals hatte ich das Gefühl eines auch für andere Dinge aufgeschlossenen Menschen. Er hatte nämlich immer einen Geigenkasten auf dem Hocker neben seinem Arbeitsplatz. — Die Zeit ist zu kurz, um noch weiter auf Einzelheiten einzugehen, und er ist ja auch schon an anderer Stelle gewürdigt worden. Aber ich möchte doch noch einmal daran erinnern, daß wir ohne KURT FELIX hier nicht zusammen wären und möchte der Hoffnung Ausdruck geben — glaube, daß sich diese auch erfüllen wird —, daß dieser kleine und so erfolgreiche und so eigenartige Kreis noch oft zusammenkommen möchte.

KLENK: Ich übergebe nun den Vorsitz Herrn Kollegen HEINZ.

HEINZ: Ich habe noch von keinem Symposium über aktiven Transport gehört, sei es, daß es in New York, Boston oder Madison abgehalten wurde, an welchem man für den einführenden Vortrag nicht Herrn USSING gebeten hätte, denn allem, was in der Ussing-Schule in Kopenhagen über aktiven Transport getan und gedacht, wird besondere Bedeutung beigemessen. Es ist immer eine wichtige Stütze bei jeder Auseinandersetzung, wenn man sich darauf berufen kann, daß das jetzt der neueste Standpunkt in Kopenhagen sei. Ich freue mich, daß auch Mosbach an der alten Tradition festhält, und ich danke Herrn USSING, daß er der Einladung gefolgt ist und bitte ihn, nun den Vortrag über: "Experimental evidence and biological significance of active transport" zu beginnen.

BIOCHEMIE DES AKTIVEN TRANSPORTS

Experimental evidence and biological significance of active transport

By

Hans H. Ussing

Institute of Biological Chemistry, University of Copenhagen, Copenhagen/Denmark

All living organisms as well as the individual cells of which they are composed are faced by two problems which are obviously antagonistic: They must retain their integrity, constraining their component molecules within a definite space, but, at the same time, require a continuous or periodic exchange of molecules with their surroundings. Off-hand, the fact that the living cells maintain a composition with respect to numerous substances which differs from that of the surroundings might be explained in one of four ways: (1) The substances cannot penetrate the cell surface (2) the substances can penetrate but the activity coefficients of the substances in the cells differs from the corresponding values in the surroundings. (3) For charged molecules the electric potential difference between cell and surroundings brings about an uneven distribution of the substances. (4) Chemical processes in the cell provides energy for an accumulation or depletion of certain substances.

It seems that all four possibilities have been exploited by living cells and the possibility listed as point four might then serve as a definitions of active transport.

A more practical definition is that a transport should be considered as active if it cannot be explained on the basis of observable physical forces. Clearly, this means that as our knowledge advances, some cases which are now listed as active transport will have to be removed from the list.

At this juncture some of you might ask what has been gained at all by introducing the term active transport. Is it more than glorified ignorance?

The answer is that once we introduce the term active transport it becomes possible to give a coherent treatment of many phenomena which hitherto seemed unrelated. Thus the concept of active Na transport leads to a unified theory for bioelectric phenomenea, regulation of cell volume, net transport of salt and water and so on. Indeed we obtain a type of simplification which is conceptually closely related to the use of activity coefficients in the physicochemical treatment of non-living systems. The activity coefficient is used to describe the deviation from "ideal behaviour". Analogously, in cases of active transport, processes in the cell surfaces or cell membranes induce a deviation from ideal behaviour. However, whereas the deviations from ideality expressed by the activity coefficient is non- directional, the deviation leading to active transport is oriented with respect to the cell structure. The activity coefficient is a scalar whereas the coefficient describing the tendency towards active transport is a vector. Parenthetically, as pointed out by KEDEM (1960) the combination of scalar and vector properties is only possible in anisotropic systems and the existence of active transport thus is an indication of a high degree of orientation of living membranes.

The advantages obtained by applying the concept of active transport is perhaps best illustrated by an example. Let us consider the distribution of potassium and sodium between a nerve fibre and the medium. Typically the cellular K concentration is many times that of the medium whereas the cellular sodium is always lower than its concentration in the medium. Of course one might ascribe this to a specific binding capacity for K and a exeptionally low "solubility" for sodium in the fibre cytoplasm. Such a view is admittedly hard to reconcile with measurements of intracelullar potentials which indicate that its potassium is usually nearly in Donnan equilibrium with the surrounding medium. However, it could be argued that the potentials measured are artefacts introduced by using liquid junctions in the micro-electrodes.

As long as we accept this "specific binding hypothesis", any change in electrolyte composition for the fibre is readily "explained" as being due to a change in binding capacity or specificity. But like other *ad hoc* hypothesis it explains everything and does not allow us to predict anything. If, on the other hand, we assume that the fibre potassium and sodium are free and their activity coefficients not materially different from those in the medium, then we are forced to accept an active transport of sodium outward to balance the inescapable leakage inward (see HODGKIN, 1957). In return for this assumption we now can make use of the beautiful theory for the electrical activity of nerve as developed by HODGKIN and HUXLEY (1952). Such quantities as threshold for electric stimulation and conduction rate as well as the detailed shape of the action potential can be predicted on the basis of a few rather simple assumptions as to the way the membrane permeabilities to K and Na change in response to changes in the membrane potential. In so far as all a scientific theory can do, even in the best of cases, is to enable us to make precise predictions, the membrane theory, including the active sodium transport concept, has demonstrated its value relative to the adsorption theory.

More direct evidence for the existance of active transport is obtained from the study of organs like intestine, kidney tubules, amphibian skin, gastric mucosa etc. in which sizable amounts of certain substances undergoe unidirectional transport. Good examples are the reabsorption of glucose in the kidney tubules which reduces the concentration in the blood ultrafiltrate to virtually nil, and the active transport of sodium from the outside to the inside of isolated frog skin, (for references see USSING et al. 1960). In these cases the transport takes place between dilute watery solutions of known composition for which reasonable estimates of the activity coefficients can be made. The sodium transport of the frog skin in particular has been studied in considerable detail. It may be appropriate to discuss in some length the evidence for the active nature of this process.

More than a hundred years ago DU BOIS-REYMOND (1848) observed that the frog skin maintains an electric potential difference between the inside and outside bathing solutions, the inside solution being normally positive relative to the outside. The conditions favouring the development of the frog skin potential were first studied by GALEOTTI (1904). Among his most important observations was that sodium or lithium ions are necessary for the maintenance of the potential. In order to explain his findings, he assumed that the diffusion coefficients for these two ions were greater for the direction outside-in than in the opposite direction. His contemporaries flatly rejected his explanation since it seemed to violate the second law of thermodynamics, but, in a way, he was not far from the truth. Numerous other attempts at explaining the potential were advanced in the following years. As late as in 1946, MEYER and BERNFELD proposed that the potential arose from the fact that hydrogen ions diffuse faster from the epithelial cells into the inside bathing solution than do bicarbonate ions, so that the formation of metabolic carbon dioxide was the real source of the potential.

In the meantime another characteristic property of the frog skin had been discovered by HUF (1935) namely that the isolated surviving frog skin, when in contact with Ringer solution on both sides, will transport chloride from the outside in. Although he did no analyse for sodium, HUF assumed the process to be an active inward transport of sodium chloride. Shortly afterwards KROGH (1937) demonstrated that chloride depleted frogs are able to take up salt (NaCl) from solutions as dilute as 10^{-5} molar. He further showed that the mechanism is specific to Na, neither K nor Ca being taken up at all (KROGH 1938). Since the blood concentration of NaCl in frogs is around 100 millimolar there can be no doubt that the frog skin performs active transport of sodium chloride. Early studies with isotopes in our laboratory indicated that even the isolated frog skin was able to perform a transport of sodium chloride against a concentration gradient. Even when the outside medium was Ringer and the inside medium 1/100 Ringer the sodium influx might exceed the sodium outflux.

A second observation which emerged from these studies was that the transport of chloride ions inward might be a consequence of the electric potential difference across the skin since the potential was usually of sufficient magnitude to raise the electrochemical potential of the chloride ion in the outside medium over that in the inside medium. Viz, $\bar{\mu}_{Cl\,(o)} > \bar{\mu}_{Cl\,(i)}$. Thus the chloride transport might be passive. By the same token, however, we have $\bar{\mu}_{Na\,(o)} \ll \bar{\mu}_{Na\,(i)}$, indicating that the inward transport of Na has to overcome the combined effects of chemical and electrical potential gradients. This illustrates the criterion for active transport proposed by ROSENBERG (1948); — Active transport is a transport against the electrochemical potential.

As far as the present example is concerned one might argue, however, that the active process were a transfer of, say, minute droplets of sodium chloride solution. If so it might not be meaningful to consider the transfer of one ionic species as an active process and the other as passive. Therefore it was necessary to work out methods for demonstrating that the cation and the anion move independently in the frog skin. Now it can easily be shown that for an ionic species passing through a membrane without interacting with other moving particles, the flux ratio is independent of the membrane structure and determined only by the difference between its electrochemical activities in the bathing solutions. Thus

$$RT\ln\,(M\ \text{in}/M\ \text{out}) = \bar{\mu}_o - \bar{\mu}_i$$

or

$$M\ \text{in}/\ M\ \text{out} = (a_o/a_i)\exp\,(FE/RT)$$

(Ussing 1949, Teorell 1949) where M in is the influx, M out the outflux $\bar{\mu}_o$ the electrochemical potential of the ion in the outside medium, $\bar{\mu}_i$ that in the inside medium. a_o and a_i, are the chemical activities of the ion in the outside and inside medium, respectively; E is the potential difference between inside and ou.side medium. R,T and z have their usual meanings. Clearly this equation describes an idealized situation since the complete absence of interaction with other moving particles is impossible. Nevertheless it turns out that the equation describes the behaviour of the chloride ion in the frog skin under all conditions (except during adrenaline stimulation which induces chloride secretion in the skin glands, Koefoed Johnsen et al. (1952). Thus the "flux analysis" gives additional evidence that the chloride transport in the skin is passive. The flux ratio for sodium, on the other hand, is often more than 100 times larger than calculated, indicating that only an insignificant amount of the sodium transport can be accounted for as passive. Interestingly enough, when the sodium transport is "uncoupled" from the metabolism by dinitro phenol the flux ratio even for this ion approaches that predicted for passive diffusion.

The flux analysis accoridng to the above mentioned simple equation has the advantage that it can be used even for systems which are not in a steady state. A behaviour in agreement with the equation strongly suggests passive transport. Deviations from the equation do not, however, necessarily indicate active transport, but only indicate strong interaction with other moving particles.

One type of interaction in particular is of interest, viz. that between solute and solvent. Thus, if the frog skin were a sieve or pore membrane through which the solution where made to flow by some mechanism, both sodium and chloride ions might be dragged along. The flow of solvent may be considered to create a "drag potential" which for each solute species is a function of the size and shape of the solvent molecules (their diffusion coefficient), the flow rate of the solvent, and the shape of the pores through which the solvent flows.

In a pore membrane with solvent flow the flux equation includes an additional term describing the drag effect:

$$\ln (\mathrm{Min}/\mathrm{Mout}) = \ln \frac{ao}{ai} + \frac{zFE}{RT} + \frac{\Delta w}{D} \int_0^{x_0} \frac{I}{A}\, \mathrm{dx}\,.$$

(KOEFOED-JOHNSEN and USSING 1951) where Δ_w is the rate of volume flow of water through unit area of the membrane, D is the diffusion constant for the substance in question. A is the fraction of unit area available to flow as a function of x, the distance from the outer surface of the membrane, and x_0 the total thickness of the membrane.

This equation shows that the drag term for a given membrane structure is primarily determined by the diffusion coefficient of the solute and the volume flow rate of the solvent.

Therefore it should be possible to estimate the drag on the sodium ion by determining the flux ratio for a test substance which has the same diffusion coefficient as Na and which is not subject to active transport.

In our experiments [USSING and ANDERSEN (1956), ANDERSEN and USSING (1957)] we used thiourea and acetamide as test substances, firstly because they have diffusion coefficients close to that of Na and secondly because it is possible to prepare two differently labeled forms of each test substance. Thus one batch of thiurea was labeled with C^{14} and the other with S^{35}. Acetamide was used in the C_1 and C_2 labelled forms. With such two forms of the same substance it is possible to measure influx and outflux simultaneously with great precision. The experiments were performed on toad skins with and without an osmotic gradient to produce solvent flow and in most experiments after treatment of the skin with antidiuretic hormone which greatly increases the rate of osmosis (presumably by increasing the effective pore diameter in some limiting layer in the skin.)

It is found that in the presence of an osmotic gradient across the skin, and particularly in the hormone treated skins, there was a large drag effect. Indeed the apparent diffusion rate in the direction of the osmosis was as much as three times that in the opposite direction. In the absence of an osmotic gradient, however, the drag effect was virtually nil.

Thus is it safe to conclude that the transport of sodium ions from the outside to the inside of amphibian skin (in the absence of an osmotic gradient) cannot be due to solvent drag. It is true that *pari passu* with the transport of sodium chloride in the frog skin there is a slight transfer of water. With Ringer on both sides it is of the order of one to two μl per cm², hour, REID (1892). Certainly this slight flow cannot in any way influence the flux ration for sodium.

Apart from the interaction with the solvent other interactions with moving particles seem to exist in living membranes. Some have the character of active transport, whereas others do not seem to require metabolic energy (exchange diffusion "single file diffusion"). For references see USSING et al. 1960.

From the foregoing it is evident that the transport of sodium in the frog skin is active and the chloride transport passive.

Moreover it would seem likely that the potential difference responsible for the chloride transport were a consequence of the inward transport of the positively changed sodium ions. A direct proof that the active sodium transport is the sole factor responsible for the electric asymmetry of the frog skin was obtained with the aid of the "short-circuiting" technique (USSING and ZERAHN (1951). The principle is that the potential drop across the skin (placed between identical solutions) is reduced to zero by an adjustable external voltage applied in series with the skin.

As long as the potential difference between inside and outside of the skin is maintained at zero, it is by definition short circuited. The current generated by the skin is equal to the current in the outer circuit and can be read on a microammeter. In this set-up the electromotive force of the skin has only to overcome the internal resistance of the skin whereas the applied external EMF overcomes the external resistance, including that of the bathing solutions.

The transport rate for sodium across the skin is so low that the chemical demonstration of the exact transport-current relationship would meet with great difficulties. If, however, the sodium influx is measured with Na^{22} and the outflux with Na^{24} (or vice versa) the net sodium transport can be obtained with great accuracy. Similarly the chloride transport can be measured by the simultaneous use of Cl^{33} and Cl^{38}.

It then turns out that during short circuit, as expected, the chloride influx and outflux are exactly equal, indicating passive behaviour. The net sodium transport, on the other hand turns out to be exactly equal to the electric current generated by the skin. The active nature of the sodium transport is apparent from the fact that it moves unidirectionally in the absence of any electrochemical gradient. Except for Li, which can partially replace Na, no other cation can support the current output of the frog skin. The short circuiting method has been applied successfully to demonstrate active sodium transport even in toad urinary bladder, toad large intestine, guinea-pig coecum and rat large intestine, (for references, see USSING et al. 1960). It has even been possible to modify the method so that it could be used to measure active sodium transport across the kidney tubule (KARGER). Even in the giant cells of the alga Halicystis ovalis it is possible to short circuit the potential between cell sap and surrounding and to demonstrate that roughly half the short circuit current comes from active sodium transport outward. The remaining current comes from active uptake of chloride ions. (BLOUNT and LEVEDAHL, 1960).

For the frog skin there thus exists an identity between sodium transport and the production of electric current. This might suggest that the active sodium transport is a major factor in determining the metabolic rate of the skin. This in fact has turned out to be correct. Thus ZERAHN (1956) has demonstrated that there exists a linear relationship between the amount of sodium transported and the amount of oxygen consumed over and above the resting consumption in the absence of sodium transport. Each sodium ion transported requires an additional consumption of $^1/_{18}$ molecule of O_2.

This relationship holds whether the sodium ions are transported between solutions of equal concentration or it is transported uphill a electrochemical potential gradient.

Thus it seems that there exists some kind of stoichiometric relationship between sodium transport and oxygen consumption.

The fact that dinitro phenol inhibits the sodium transport might indicate that the transport depends on the consumption of energy rich phosphate bonds. More direct evidence for the consumption of ATP in active ion transport has been obtained from work on red cell hosts (GARDOS, 1954) and chephalopod giant axons. (HODGKINS and KEYNES, 1955).

The demonstration of the direct and stoichiometric relationship between transport and metabolic process is obviously a very satisfactory method for demonstrating active transport. It requires, however that the transport in question consumes a major fraction of the total metabolic energy of the system. This is the case for the sodium transport of the amphibian skin. Also the metabolic rate of the kidney tubules is determined primarily by the amount of sodium transported (LASSEN et al.).

This relationship can be clearly demonstrated in experiments with kidney slices. (LASSEN and THAYSEN. 1961).

Similarly, the hydrogen ion excretion of the gastric mucosa is a primary factor in determining its metabolic rate (DAVENPORT, 1957, DAVIES, 1957).

In many cases, however, the energy involved in transporting a certain molecular species may be quite insignificant compared with the total metabolism of a given cell or organ.

In such cases the other methods mentioned above must be used for identifying the active nature of the transport.

To summarize, these methods are:

1) Transport against the electrochemical potential gradient.

2) A flux ratio which is incompatible with passive transport. This method is safe for transport uphill and between identical solutions but certain types of interaction (single file diffusion) may stimulate active transport if the process is going downhill the electrochemical potential gradient. Only one such case has been observed so far.

3) Identity of transport and electric current output.

4) Proportionally between transport and extra oxygen uptake.

5) Inhibition by metabolic inhibitors.

This last method is not always safe since the inhibitor may also change conditions for passive transport for instance by changing the membrane structure.

The question as to how the active transports are brought about is another matter. So far, to my knowledge, no active transport has been cleared up in detail. The fact remains, however, that active transport exists. It plays an enormous role in all living cells and presents a fascinating challenge to biochemists and biophysicists.

References

ANDERSEN, B., and H. H. USSING: Solvent drog on non-electrolytes during osmotic flow through isolated toad skin and its response to antidiuretic hormone. Acta physiol. scand. **39**, 228—239 (1957).

BLOUNT, R. W., and B. H. LEVEDAHL: Active sodium and chloride transport in the single celled marine alga Halicystis Ovalis. Acta physiol. scand. **49**, 1—9 (1960).

DAVENPORT, H. W.: Metabolic aspects of gastric acid secretion. Metabolic aspects of transport across membranes. p. 295—302. E. D. Murphy, Q. R. University of Wisconsin Press 1957.

DAVIES, R. E.: Gastric hydrochloric acid production — the present position. Metabolic aspects of transport across cell membranes. p. 277—293. Ed. Murphy, Q. R. University of Wisconsin Press 1957.

DU BOIS-REYMOND, E.: Untersuchungen über tierische Elektricität. Berlin 1848.

GALEOTTI, G.: Concerning the EMF which is generated at the surface of animal membranes on contact with different electrolytes. Z. physik. Chem. **49**, 542—562 (1904).

GARDOS, G.: Akkumulation der Kaliumionen durch menschliche Blutkörperchen. Acta physiol. Acad. Sci. hung. **6**, 191—199 (1954).

HODGKIN, A. L., and A. F. HUXLEY: A quantitative description of membrane current and its application to conduction and excitation in nerve. J. Physiol. (Lond.) **117**, 500—544 (1952).

HODGKIN, A. L., and R. D. KEYNES: Active transport of cations in giant axons from Sepia and Loligo. J. Physiol. (Lond.) **128**, 28—60 (1955).

HODGKIN, A. L.: The ionic basis of electrical activity in nerve and muscle. Biol. Rev. **26**, 339—409 (1957).

HUF, E.: Versuche über den Zusammenhang zwischen Stoffwechsel, Potentialbildung und Funktion der Froschhaut. Pflügers Arch. ges. Physiol. **235**, 655—673 (1935).

KARGER: in preparation.

KEDEM, O.: Criteria of active transport. Membrane Transport and Metabolism-Symposium. Prag, 1960 in press.

KOEFOED-JOHNSEN, V., H. H. USSING and K. ZERAHN: The origin of the short-circuit current in the adrenalin stimulated frog skin. Acta physiol. scand. **27**, 38—48 (1952).

KROGH, A.: Osmotic regulation in the frog (R. esculenta) by active absorption of chloride ions. Skand. Arch. Physiol. **76**, 60—74 (1937).

KROGH, A.: The active absorption of ions in some freshwater animals. Z. vergl. Physiol. **25**, 335—350 (1938).

LASSEN, U. V., and J. H. THAYSEN: Correlation between sodium transport and oxygen consumption in isolated renal tissue. Biochem. biophys. Acta. **47**, 616—618 (1961).

LASSEN, N., O. MUNCH and J. H. THAYSEN: In press.

MEYER, K. H., and P. BERNFELD: The potentiometric analysis of membrane structure and its application to living membranes. J. gen. Physiol. **29**, 353—378 (1946).

REID, E. W.: Report on experiments upon "Absorption without Osmosis". Brit. med. J. **1892**, 323—326.

ROSENBERG, T.: The concept and definition of active transport. Symp. Soc. exp. Biol. **8**, 27—41 (1954).

TEORELL, T.: Membrane electrophoresis in relation to bio-electrical polarization effects. Arch. Sci. Physiol. **3**, 205—219 (1949).

USSING, H. H.: The distinction by means of tracers between active transport and diffusion. Acta physiol. scand. **19**, 43—56 (1949).

USSING, H. H., and B. ANDERSEN: The relation between solvent drag and active transport of ions. Proceedings of the Third International Congress of Biochemistry, Brussels 1955. Academic New York: Press 1956.

USSING, H. H., P. KRUHØFFER, J. HESS TAYSEN and N. S. THORN: The alkali metal ions in biology. Handbuch der experimentellen Pharmakologie 13 Bd. (1960).

USSING, H. H., and K. ZERAHN: Active transport of sodium as the source of electric current in the short-circuited isolated frog skin. Acta physiol. scand. **23** Fasc. 2—3 (1951).

ZERAHN, K.: Oxygen consumption and active sodium transport in the isolated and short circuited frog skin. Acta physiol. scand. **36**, 300—318 (1956).

Diskussion

Diskussionsleiter: HEINZ, *Frankfurt/M.*

Ich danke Herrn U. für den anschaulichen Überblick, den er uns über die Grundlagen des aktiven Transportes gegeben hat. Ich nehme an, daß Herr U. gern einige Fragen dazu beantworten wird, und ich eröffne die Diskussion.

BADER (München): Herr Professor, Sie haben über den aktiven Transport von Na durch eine Zellage gesprochen, d. h. das Na muß durch zwei Membranen, eine äußere und eine innere. Findet der aktive Transport in beiden Membranen statt oder ist es nicht so, daß beim Einflux in einer Membran der Transport passiv ist, in der anderen aber aktiv? Wie verhält es sich beim Ausflux? Ist er nicht auch einmal passiv und einmal aktiv?

USSING: The question just raised, can be answered by studying the nature of the electrical potential, the electromotive force of the frog skin system.

If the skin is treated with a trace of Cu^{++} which lowers the Cl^- permeability or if the chloride is replaced by sulphate that does not penetrate — we get a system where there is no net transport of salt but only development of an electromotive force. In this system it is possible to study separately the reaction of the outside and inside membranes to sodium and potassium and from such studies certain conclusion can be drawn as to the site of the active transport. The skin reacts to changes in ionic composition of the media as if the inward facing membrane were a potassium electrode that is freely permeable to potassium but impermeable for sodium. Thus when the "inside" potassium is increased the potential drops with a slope which is very close to that of an ideal potassium electrode.

And to prove, that this indeed is so: that the inner membrane is permeable to potassium and not to sodium, we have measured the thickness of the epithelium under the microscope while changing the bathing solutions. It turns out that when sodium chloride is replaced by potassium chloride in the outside solution, potassium does not go into the skin. The outer membrane is impermeable to K, but somewhat permeable to Na. The volume of the epithelium does not change. But if potassium replaces sodium on the inside with the chloride ion as the anion you see the epithelium swell from 60 to 90 μ. If you go back to normal Ringer solution it shrinks again, in other words the inward facing is permeable to K^+ but impermeable to Na^+.

If however you do similar changes of solution on the outside of the skin it turns out that the potential now behaves as if the outer membrane were a sodium electrode. That is, when the Na^+ concentrations are varied between 1 to 100 milliomolar the potential increases in a linear fashion as if the outer membranes were permeable to sodium ions but not to K^+ or the anion which in this case was $SO_4^{=}$; but Cl^- as anion gives the same results for Cu^{++} treated skins. So we then have the following picture:

It looks as if the potential were the sum of two diffusion potentials: the Na^+ potential at the outer membrane and the K^+ potential at the inner membrane. Thus it would seem that we don't need any active transport to explain our observations. This in fact is true as long as the anion can not penetrate. But now let us look at a model of the cell or rather of the epithelium: the outside medium has high sodium and sulfate. Na^+ diffuses through the outer membrane and gives rise to the first diffusion potential. The cell has a high K^+ concentration. K^+ cannot pass the outer membrane but diffuses through the inner membrane and gives rise to the second diffusion potential. No net transfers of salts take place because the anions are non-diffusible. But now let us see, what happens when sulfate is replaced by Cl^-. Then Cl^- diffuses through the whole epithelium and you will see that the battery runs down. The cell would fill up with NaCl and K^+ would run out and finally you loose the potential. So the situation is only stable if the Na^+ that gets in is also pumped out at the inward facing membrane. The Na^+ that diffuses into the cells has to be lifted up to at least the same electrochemical potential as it has in the inside bathing solution in order for the situation to remain static. This argument places the active Na^+ transport at the inner membrane of the skin. There are some additional observations which seem to indicate that it is not a pure sodium pump that in fact it returns K^+. I will only mention one piece of evidence viz that in the absence of K^+ in the inner solution there is no Na^+ transport. Thus it looks as if K^+ is the partner in the transport. The sodium is being pumped from cell to inside solution and K^+ is being returned to the cell and diffuses back to where it came from. K^+ is moving in a closed circle and only sodium undergoes net transport. You can see that it is only in the presence of a diffusable anion that there is evidence for net transport. This also is evident if you measure the oxygen consumption. If there is no Na^+ outside or if the membrane is tight to salt (in the presence of SO_4^{--} for instance) the O_2-consumption of the skin is very low. If the anion is allowed to follow or if the skin is short circuited so that there is a net transport of Na^+ through the system then the O_2-consumption goes up. Only when there is a net transport of Na^+ through the pump you measure an increased O_2-consumption. There seems to be no difference in principle between the subcellular elements necessary to bring about net transport of salt through an epithelium like the frog skin and those maintaining the ione composition of cells. Let us consider the muscle or the nerve fiber. It has its Na^+ pump which keeps the Na^+ low and the K^+ high in the cell. The only difference is that in the polar cells of the frog skin the pump is absent on the outside or if not completely absent it is working at very slow rate. Instead the outer membrane is Na^+ selective, so that Na^+ can diffuse in. At the inward facing membrane Na^+ is being pumped out just as in any other cell. Thus we have a practically uniform picture of how the Na^+ pump works in individual cells and how it works in an organized epithelium.

Bücher (Marburg): Sie haben einen Term angegeben, der den Fluß des Wassers berücksichtigt. Meine Frage ist, wie es sich hier mit dem Parameter A verhält oder ob es sich um freie oder gehemmte Diffusion handelt? Hängt A auch ab von der Natur Ihres etikettierten Stoffes?

Ussing: This is of course a tricky point of the treatment, because it requires that we have to assume that all the pores are identical. It is allright as long as all the pores are of the same shape and that the only way that any substance gets through the membranes is through the pores. However if there are other pores, say very narrow ones which allow water but not the ion in question, then we are in trouble, because some of the waterflow is not concerned with the process we are dealing with. Only in a system with pores of the same shape and forms does this equation hold. But to be careful we used as one of our test substances H_2O itself, that is labelled water. Now this equation indicates, that if this is the slope of the curve for thiourea then the slope for H_2O should be so much less, inversely proportional to the diffusion coefficient. The ratio of the diffusion coefficients is 1:2. The ratio of the slopes were very close to the ratio 1:2. In other words: this treatment seems to indicate that the path which is used by water is the same as that used by thiourea and acetamide. Of course you could still claim that Na^+ uses a different path. I could not deny that. It only seems that the path which is used by water is the important one in this treatment. Because if Na^+ goes by a different route, then the water flow does not act on

Na^+ and then there would be no drag term at all. Thus I think it is reasonable safe to use the present treatment. If there is any drag effect on Na^+ it can not be larger than the one described by the above equation.

ULRICH (Göttingen): Welche Methode halten Sie für die beste zur Messung der elektromotorischen Kraft e_m der Natriumpumpe?

USSING: It is of course a little difficult to answer this question. Ideally if one could get a situation where the anion did not move at all and the Na^+ was still transported in a satisfactory way this probably should approach the max. electromotive force. In some cases we had luck with $SO_4^=$ as anion. In some skins you get potentials as high as 160 mV. No other method gives higher values. But the trouble is that not all skins can be forced to give maximum potentials in sulphate. It probably is because some of them are leaky to other ions and soon as just one species leaks, you get in turn a short-circuit. You can never be sure, if you take a certain skin, wether or not some ionic species say H^+, PO_4^{--} or SO_4^{--} passes through one of the membranes. So the best you can do is to make a number of experiments and see what is the max. value.

HEINZ: Ich schlage vor, daß wir vor weiteren Fragen noch den nächsten Vortrag abwarten, möglicherweise werden sie dadurch schon beantwortet. In dem Vortrag von Professor NETTER über mögliche Mechanismen und Modelle für aktive Transportvorgänge werden sicher einige Probleme behandelt, die eng mit den Fragen an Herrn USSING zusammenhängen.

Mögliche Mechanismen und Modelle für aktive Transportvorgänge*

Von

Hans Netter

Institut für physiologische Chemie und Physikochemie, Kiel

Mit 16 Textabbildungen

Einleitung

Der Grund, weswegen im Folgenden nur von den Möglichkeiten des Zustandekommens aktiver Transportvorgänge gesprochen werden kann, ist einfach anzugeben: ihr Mechanismus ist noch unbekannt. Ihn kennen zu lernen, ist ein dringendes Anliegen der kausalen Biologie, da die Existenz und die Leistungen der Zellen und Gewebe von seinem Funktionieren abhängen. Der aktive Transport selbst aber ist ein biochemischer Vorgang, der allerdings zunächst als physikalisch-chemische Größe, nämlich als Konzentrationsänderung meßbar wird. Eine als aktiver Transport zu bezeichnende Veränderung von Konzentrationen gegenüber den bei passiver Verteilung zu erwartenden ist wohl eine der einfachsten Äußerungen einer Wechselwirkung zwischen der statischen Feinstruktur der Zelle und den sich an ihr abspielenden chemischen Vorgängen.

Bei den folgenden Erörterungen soll nur auf den sich unmittelbar mit einem Stoff und an einer definierten Struktur vollziehenden Vorgang eingegangen werden, ohne die weiteren Folgen für das biochemische Geschehen zu beachten. Es soll hier auch nicht auf die oft beobachteten und schon lange diskutierten physiologisch bedingten oder experimentell erzeugten Änderungen einer passiven Durchlässigkeit Rücksicht genommen werden, so wichtig dieses Phänomen und seine chemische Grundlage auch sein mögen.

In diesem Zusammenhang muß auch die Frage unerörtert bleiben, ob der mikromorphologische Zellaufbau besser als Sorptions-,

* Herrn Professor H. H. Weber zum 65. Geburtstag gewidmet.

also Schwamm- oder als Ballon-, also Membranstruktur zu beschreiben ist oder mehr in der einen oder anderen dieser beiden Formen wirksam wird. Sicher ist aber, daß bei den am genauesten untersuchten Objekten, nämlich den Erythrocyten, den marklosen Nerven und der Hefe, die nun einmal vorhandene Membranstruktur auch funktionell vollständig im Vordergrund steht.

I. Das Problem der Koppelung physikalischer Flüsse

Der aktive Transport ist ein endergonischer, d. h. Energie verbrauchender Konzentrierungsvorgang. Dabei findet die Konzentrierung nicht durch Neubildung eines Stoffes, sondern durch seine Beförderung von einer Stelle zu einer anderen statt. Der Vorgang ist also erstens *richtungsbedingt*. Da er aber zweitens energieabhängig und als Teilprozeß *unfreiwillig* ist, wird sein Wesen darin bestehen müssen, daß ein sich freiwillig in einer Richtung vollziehender Stoffdurchtritt eine zweite Strömung aktiv, d. h. unter Energieaufnahme aus dem ersten Vorgang, in Gang setzt. Für die einzelnen Stoffe liegt dann immer je ein Ausfluß (outflux) und ein Einfluß (influx) vor, niemals ein statisches Gleichgewicht, wohl aber dann ein steady-state, wenn für einen Stoff beide Flüsse entgegengesetzt gleich sind, der sog. Nettotransport also Null ist.

Die Schaffung oder Aufrechterhaltung eines solchen dynamischen Gleichgewichtes außerhalb des statischen erfordert die dauernde Zuführung von Energie, d. h. eine Energieströmung, welche an eine Stoffströmung gebunden ist und andere hervorrufen kann. Diese können schon unmittelbar den unfreiwilligen Flux darstellen, oder er ist erst sekundär mit ihnen gekoppelt. Prinzipiell sollten daher der *treibende und der bewegte Strom* in physikalischer und chemischer Beziehung bekannt sein, wenn man die *Koppelung* beider Flüsse untersuchen will. Ihren Mechanismus zu erkennen, ist das Hauptproblem der Erforschung des aktiven Transportes.

Flüsse werden als die in der Zeiteinheit durch die Einheit des Querschnittes hindurchtretenden Stoff-, Wärme-, Elektrizitätsmengen usw. definiert. Ihre Größe ergibt sich als Produkt einer Triebkraft mit einem Leitfähigkeits-, Durchgangs-, Fluß-Koeffizienten usw. Die Triebkraft ist als Gradient der entsprechenden Potentiale in der jeweiligen Flußrichtung anzugeben, wie der Spannungsgradient der elektrischen Energie in Volt, also als Potentialabfall. Er ergibt multipliziert mit dem Koeffizienten der Leitfähigkeit, d. h. dem reziproken Wert des gegebenen Widerstandes die Stromstärke, also den Fluß der Elektrizitätsmengen durch den Querschnitt pro Zeiteinheit.

$$J_e = \frac{E}{W} = L_e \cdot E; \text{ bzw. allgemein: } J_i = L_i \cdot X_i\,. \qquad (1)$$

Die *Verknüpfung* solcher Ströme wird *formal durch* Einführung der gekreuzten oder *Koppelungskoeffizienten* vorgenommen. Ihre Zahl ergibt sich aus der der miteinander zu verbindenden Ströme (n) zu n^2. Es sollen sich z. B. der elektrische Strom (I) und der Flüssigkeitsstrom (J) beim Durchtritt durch eine beide miteinander koppelnde Membran beeinflussen. Dabei entsteht der elektrische Strom durch ein angelegtes Spannungsgefälle X_e und der Flüssigkeitsstrom durch einen mechanischen Druckunterschied X_p. Die Ströme sind nun wechselseitig von den entsprechenden Druck- bzw. Potentialdifferenzen abhängig. Die Größe dieser Abhängigkeit findet in dem Wert der Koppelungskoeffizienten ihren Ausdruck:

$$I = L_{ee} \cdot X_e + L_{ep} \cdot X_p \qquad J = L_{pe} \cdot X_e + L_{pp} \cdot X_p\,. \qquad (2)$$

Die 2 Gleichungen enthalten 4, d. h. n^2 Koeffizienten, von denen die gleich indizierten (L_{ee} und L_{pp}) mit den jeweiligen Leitfähigkeits- bzw. Wasserdurchgangs-Konstanten identisch sind. Die Koeffizienten mit gekreuzten Indices dagegen drücken die Strenge der Verknüpfung zwischen beiden Flüssen aus. Für sie gilt das Gesetz der Vertauschbarkeit der Indices (Onsager), d. h. $L_{ep} = L_{pe}$ usw.

Dadurch wird die Zahl der Koeffizienten auf $\frac{n(n+1)}{2}$ reduziert, hier also auf 3. Bei 3 Gleichungen mit je 3 Gliedern, welche etwa einen Diffusionsstrom (d), einen Flüssigkeitsstrom (o) und einen chemisch getriebenen Reaktionsstrom (r) verknüpfen, sind nicht 9, sondern nur 6 Faktoren für die Beschreibung erforderlich.

Mit Hilfe obiger Gleichungen lassen sich nun die gekoppelten Phänomene der elektrischen und osmotischen Strömung und die der Erzeugung elektrischer Spannungen durch Druck und der von Druck durch elektrische Spannungen erfassen und durch die Koeffizienten quantitativ beschreiben. Es lassen sich den Gleichungen (2) durch Nullsetzen der Flüsse leicht folgende Beziehungen entnehmen: für die *Strömungspotentiale*, die bei Druckänderungen ohne primären elektrischen Stromfluß ($X = 0$) entstehen (Strömungspotential):

$$-\left[\frac{X_e}{X_p}\right]_{I=0} = \frac{L_{ep}}{L_{ee}}, \tag{3a}$$

für die Druckänderung durch elektrische Potentiale ohne vorherigen Flüssigkeitsstrom ($J = 0$) *(Elektroosmotischer Druck):*

$$-\left[\frac{X_p}{X_e}\right]_{J=0} = \frac{L_{pe}}{L_{pp}}, \tag{3b}$$

für die Stromentstehung bei mechanischer Druckdifferenz unter elektrischem Kurzschluß ($X_e = 0$) *(Strömungsströme):*

$$\left[\frac{I}{J}\right]_{X_e=0} = \frac{L_{ep}}{L_{pp}}, \tag{3c}$$

für die osmotischen Ströme beim elektrischen Stromfluß ohne anfängliche Druckdifferenz ($X_p = 0$) *(Elektroosmose):*

$$\left[\frac{J}{I}\right]_{X_p=0} = \frac{L_{pe}}{L_{ee}}. \tag{3d}$$

Die Eliminierung der Koeffizientenbrüche ergibt folgende quantitative Beziehung zwischen den entsprechenden gekuppelten Erscheinungen:

$$\left[\frac{X_p}{X_e}\right]_{J=0} = -\left[\frac{I}{J}\right]_{X_e=0}, \tag{4a}$$

el. osmotischer Druck Strömungsstrom

$$\left[\frac{X_e}{X_p}\right]_{I=0} = -\left[\frac{J}{I}\right]_{X_p=0}. \tag{4b}$$

Strömungspotential el. osmotischer Fluß

Das Experiment steht in Übereinstimmung mit obigen Ableitungen; es ist z. B. die erzeugte Feldstärke nach (3a) dem angelegten mechanischen Druck proportional usw. Die exakte Umkehrbarkeit der Vorgänge, welche mit einer mechanischen und elektrischen Strömung verknüpft sind, beweist die Reziprozität irreversibler Prozesse in geeigneten Systemen. Dabei ist die Erzeugung einer Flüssigkeitsströmung entgegen einem mechanischen Druck durch einen fließenden Strom usw. ein gutes *Lehrbeispiel* für die Verknüpfung von Strömungen, welche im *aktiven Transport* zum Ausdruck kommt. Die zugrunde liegende Koppelung vollzieht sich in den Membranporen an der Stelle, wo der festhaftende Teil der elektrischen Doppelschicht sich von dem beweglichen trennt, und wo die angelegte Spannung eine Verschiebung der geladenen Flüssigkeitsfäden hervorruft, während umgekehrt eine mechanische Bewegung des Wassers im Druckgefälle einen potentialschaffenden Transport von geladenen Teilchen und damit auch einen elektrischen Strom hervorruft (Lit. vgl.[1]) (Abb. 1).

So einfach liegen nun die Verhältnisse bei den uns interessierenden aktiven Transportvorgängen, d. h. Fluxkuppelungen über die Zellmembranen nicht. Das ist vor allem schon deswegen der Fall, weil die Zahl der Partner immer größer zu sein pflegt als im Elektroosmose-Beispiel. Allein das Dazwischentreten chemischer Reaktionen, welche einen gerichteten Stoffstrom hervorrufen, erfordert besondere Beachtung. Diesen Flux aber richtig einzusetzen, wird dann besonders schwierig, wenn die chemischen Stoffumsetzungen in ihrem Vorkommen auf die sehr dünne und kaum noch als Phase anzusprechende Membran beschränkt sind. Dennoch gibt es einige den biologischen Verhältnissen näherstehende Kombinationen, die eine einfachere Analyse erlauben.

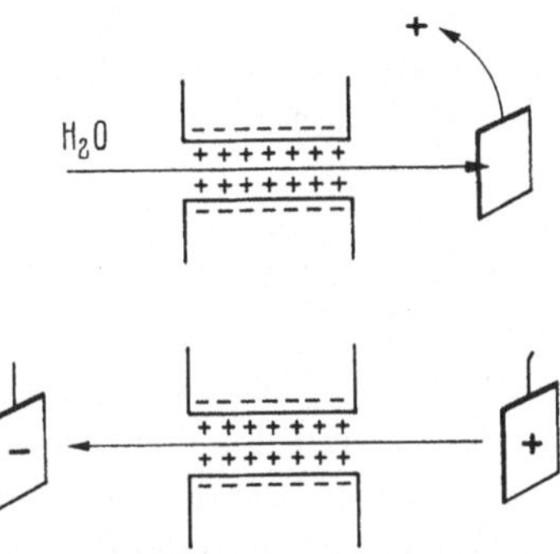

Abb. 1. *Elektroosmose (Strömungsstrom).* Die obere Pore gibt die elektrische Stromrichtung bei mechanischer Flüssigkeitsbewegung (Strömungsstrom), die untere die Richtung der elektroosmotischen Wasserbewegung im elektrischen Feld an

Betrachtet man nur den *Diffusionsfluß* (J_d) einer Substanz und den *osmotischen* des Lösungsmittels (J_o) nebst einem energieumsetzenden (J_r) von Partnern einer chemischen *Reaktion,* dann ergeben sich im einfachsten Fall 3 Gleichungen mit insgesamt 9 Gliedern (Gl. 5). Wir richten unser Augenmerk auf die Werte der Koeffizienten mit gekreuzten Indices. Beachten wir zunächst nur den Diffusionsflux J_d, dann wird, wenn der Koeffizient L_{do} den Wert Null annimmt, damit zum Ausdruck gebracht, daß eine osmotische Strömung des Lösungsmittels durch die Membran den Diffusionsvorgang des betrachteten Stoffes d nicht beeinflußt. Andernfalls existiert ein solcher Einfluß, von USSING (KOEFOED-JOHNSEN u. USSING, 1953) untersucht und "solvent drag" genannt. Über seine Richtung gibt das Vorzeichen von X_o Auskunft. Wichtig für unser Problem ist L_{dr}. Ist seine Größe von 0 abweichend, dann existiert ein Einfluß des Reaktionsablaufes auf den Diffusionsstrom von d. Ob er ihn vergrößert, verkleinert oder umkehrt, hängt von der Größe und der Richtung der chemischen Triebkraft (X_r, Differenz der chemischen, bzw. elektrochemischen Potentiale) und des Koppelungsfaktors (L_{rd}) ab. Er ist eine Vektorgröße. Formal entsteht auf diese Weise ein chemisch getriebener gerichteter Fluß. Wenn L_{or} verschwindet, wird zum Ausdruck gebracht, daß kein aktiver, d. h. chemisch getriebener Transport von H_2O vorhanden ist. Nur bei reellen Werten von L_{or} besteht zwischen der Wasserbewegung und der Energielieferung eine direkte Koppelung. Wenn außerdem $L_{dr} = 0$ ist, kann bei $L_{or} \neq 0$ und $L_{do} \neq 0$ noch ein aktiver, aber nun indirekter Stofftransport auf dem Wege über *r—d—o* stattfinden.

$$\begin{array}{l|l} J_d = L_{dd} \cdot X_d + L_{dr} \cdot X_r + L_{do} \cdot X_o & L_{dr} = L_{rd} \\ J_r = L_{rd} \cdot X_d + L_{rr} \cdot X_r + L_{ro} \cdot X_o & L_{do} = L_{od} \\ J_o = L_{od} \cdot X_d + L_{or} \cdot X_r + L_{oo} \cdot X_o & L_{or} = L_{ro} \end{array} \qquad (5)$$

Ihn bei den einzelnen Schritten in vivo zu verfolgen, ist z. Z. praktisch noch nicht durchführbar.

II. Möglichkeiten und Modelle

A. Mit ruhender Membran. Ganz allgemein ist bei unseren augenblicklichen Kenntnissen über die quantitativen Verhältnisse beim aktiven Transport eine Gewinnung von Kreuzkoeffizienten unmöglich und das Operieren mit ihnen in speziellen Fällen daher nicht sinnvoll. Man hat sich deshalb oft mit dem Studium von mehr oder weniger gut geeigneten Modellen befassen müssen. Eines der einfachsten wurde von HARTLEY 1937 diskutiert. Nach einer Seite wird eine Zelle durch eine grobe, weitgehend permeable Membran von außen abgetrennt. Ihr Raum wird durch eine semipermeable Membran so unterteilt, daß der von außen abgewandte Teil einen

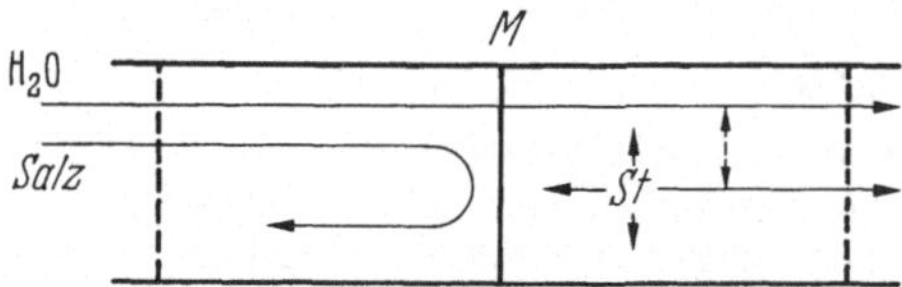

Abb. 2. Hartley-Modell (1937)

Stoff (St) dauernd in großer Menge produziert; er wird osmotisch Wasser über das andere Kompartiment von außen aufnehmen und die in diesem Teil vorhandenen Stoffe konzentrieren, soweit sie nicht zurückdiffundieren. So wird ein einseitiger Wasserstrom durch die Reaktion aufrechterhalten (Abb. 2).

Ersichtlicherweise sind hier die 3 gezeichneten Ströme nicht unabhängig voneinander. Jedoch zeigt das Modell weiter keine näheren Beziehungen zu bekannten cellularen Transportvorgängen außer der allgemeinen, daß alle Flußvorgänge durch eine chemische

Reaktion in Gang gesetzt werden und das System nur im Fließzustand existent ist.

Ohne zunächst eine chemische Reaktion einzusetzen, gelang es W. KUHN[4] in Modellversuchen durch *Kombination unterschiedlich durchgängiger Membranen* Stoffkonzentrierungen zu realisieren. Hierbei wird die zur Konzentrierung benötigte *Energie aus der osmotischen Arbeit bei der Verdünnung* konzentrierterer oder gleichstarker Lösungen eines anderen Stoffes gewonnen. Im Versuch wurden Rohrzuckerlösungen mit Hilfe von Phenollösungen durch Wasserentzug über nur für Wasser und nur für Phenol durchgängige Membranen konzentriert. Hierbei dringt Phenol in die Zuckerlösung ein, welche nach der einen Seite durch die semipermeable Membran von weiterer Zuckerlösung getrennt ist. Die Konzentrierung geschieht dadurch, daß der Inhalt des Mittelraumes durch das von der anderen Seite übergetretene Phenol einen höheren osmotischen Druck gewinnt. Der reinen, auf der anderen Seite vorhandenen und durch eine nur für Wasser durchgängige Membran abgetrennten Zuckerlösung wird auf diese Weise Wasser entzogen, die Lösung also konzentriert (Abb. 3). Ob Glycerylphosphorylcholin in der Niere die Rolle der Modellsubstanz „Phenol" übernimmt, wie es BÜCHER[6] auf Grund ihres Verteilungsverhaltens in den einzelnen Markabschnitten diskutiert, ist z. Z. noch nicht zu entscheiden.

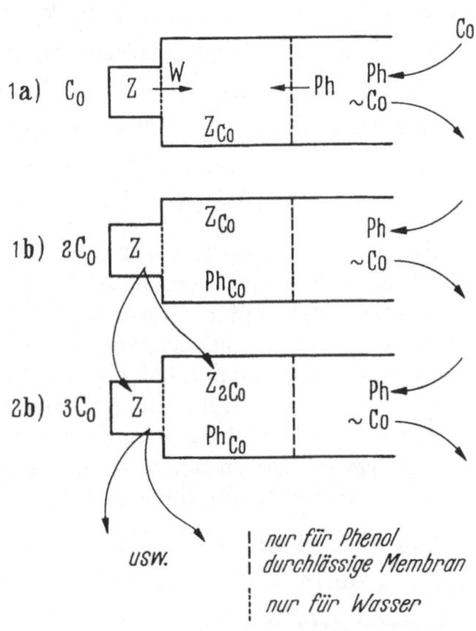

Abb. 3. Konzentrierung einer Rohrzuckerlösung mit einer gleich konzentrierten Phenollösung nach KUHN und RYFFEL. *Z* Zuckerlösung, *Ph* Phenollösung. Ausgangskonzentration beider ist c_0. Wasser wird osmotisch aus *Z* bis zur Erreichung von $2c_0$ entfernt; das gleiche geschieht beim nächsten Schritt mit der so angereicherten Lösung im linken Raum, nachdem sie in die beiden linken übertragen und wieder von rechts mit *Ph* durchströmt wurde. So wird $3c_0$ für die reine Zuckerlösung erreicht usw.

Eine Potenzierung des Effektes ist durch eine Gegenstromanordnung möglich. Eine solche Anordnung wird auch als *Haar-*

nadelgegenstromprinzip bei einer osmotischen Konzentrierung von KUHN und HARGITAY (1951)[5] benutzt, welche nach Analogie zum *osmotischen Stempel* einen *mechanischen Druck zur Konzentrierung* einsetzt. Die Kraft, welche zum Abpressen des Lösungsmittels aus einer durch eine semipermeable Membran abgetrennten Lösung erforderlich ist, kann dann sehr klein gehalten werden, wenn wiederum durch ein Gegenstromprinzip eine *Multiplikation der Teileffekte* an den Einzel-Membranelementen erreicht wird. Beachtliche Konzentrierungen können auf diese Weise aber nur dann erhalten werden, wenn kleine Konzentrationsleistungen von den Zellen auf andere Weise vollzogen werden.

KUHN erklärt nach einem weitgehend anerkannten Vorschlag die Harnkonzentrierung in den Nierenausführungsgängen nach diesem Haarnadelgegenstromprinzip unter Verwendung der Teilleistungen der einzelnen Epithelzellen im Sinne eines aktiven Transportes. Zum Beweise für sein Vorkommen an diesen Stellen hat ULLRICH[6] bedeutende Beiträge geliefert. Die Kombination des Multiplikationsprinzips mit der aktiven Zelleistung scheint eine befriedigende Erklärung für die abschließende Konzentrierung des Urins in der Säugetierniere zu geben; sie *setzt* aber dabei die Existenz — wie bereits betont — eines *aktiven cellulären Stofftransportes voraus.*

Ein Modell für die auffällige *Anreicherung des Kaliums in der Zelle* wurde frühzeitig (NETTER, 1928)[7] aufgestellt und in Modellversuchen geprüft. Hierbei wurde eine kontinuierliche Bildung von Säuren in der Zelle und eine selektiv kationenpermeable Membran vorausgesetzt. Die Modellversuche sahen zunächst an den selektiv für H^+- und K^+-Ionen durchgängigen Membranen die Einstellung eines partiellen Ionengleichgewichtes vor, bei dem die elektrochemischen Potentiale der Elektrolyte zwar beiderseits nicht gleich sind, bei denen aber die permeablen Ionenpaare der Gleichgewichtsbedingung: $K_i \cdot H_a = K_a \cdot H_i$ gehorchen (Abb. 4).

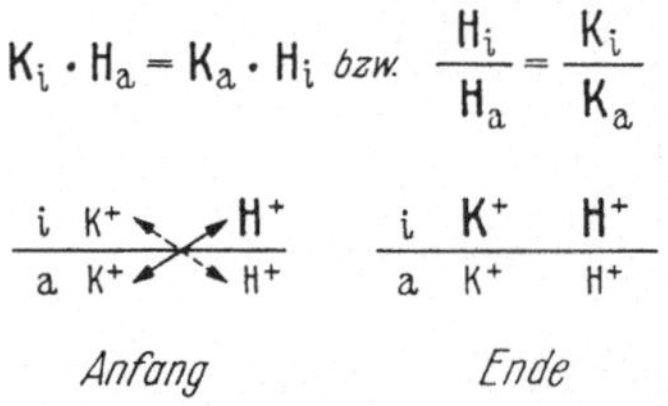

$$K_i \cdot H_a = K_a \cdot H_i \quad \text{bzw.} \quad \frac{H_i}{H_a} = \frac{K_i}{K_a}$$

	Anfang			Ende	
i	K^+ ↘	↗ H^+	i	K^+	H^+
a	K^+ ↗	↘ H^+	a	K^+	H^+

Abb. 4. Partielles Ionengleichgewicht für selektiv kationenpermeable Membranen

Man erhält durch den Tausch von H_i gegen K_a eine erhebliche, d. h. bis 40fache Anreicherung der K-Ionen auf der Seite der Membran, welche auch die höhere Konzentration an H^+-Ionen besitzt.

Diese Folgerung ist auch aus dem echten *Donnangleichgewicht* für Membranen abzuleiten, welche auch noch Anionen und damit Elektrolyte passieren lassen können, wie es für viele Zellmembranen anzunehmen ist. Sie konnte jedoch nicht, oder mindestens nicht generell, bestätigt werden. Denn eine etwa 50fach höhere H^+-Ionen-Konzentration im Innern der Zelle gegenüber außen, welche nach dem gleichen Quotienten der K^+-Ionenkonzentration zu erwarten wäre, wird praktisch nicht gefunden. Für die Abweichung ergeben sich 2 Möglichkeiten: 1. Das Bestehen eines Scheingleichgewichts, bei dem die Forderung nicht für den ganzen Zellraum gilt, sondern nur in unmittelbarer Nähe der Membran erfüllt und damit kaum nachweisbar ist. 2. Es könnte ein Steady-State bestehen, in dem H^+-Ionen aus der unmittelbaren Nähe der Membran durch aktiven Transport ständig abgeführt werden (KOSTYUK und SOROKINA 1960)[8].

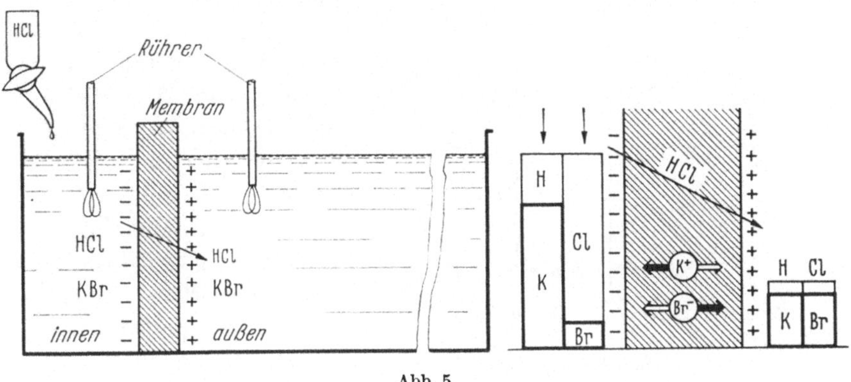

Abb. 5

T. TEORELL (1937)[9] hat Anreicherungen von Kalium-Ionen in Systemen erreicht, bei denen die Einstellung von *partiellen Gleichgewichten* noch auffälliger als in den alten Versuchen ist. Er ließ z. B. Salzsäure in einem starken Konzentrationsgefälle durch poröse Membranen nach außen diffundieren; dabei zeigen auf beiden Seiten in gleicher Ausgangskonzentration zugesetzte, ebenfalls frei diffusible Kaliumsalze eine Verteilung nach dem Donnangleichgewicht: es reichern sich K^+-Ionen auf der Seite der Salzsäure an und Cl- oder Nitrationen auf der Gegenseite. Diese *durch den Diffusionsstrom erzwungene Verteilung* bleibt solange bestehen, wie das Konzentratinosgefälle für die Säure aufrechterhalten wird. Die Verteilung selbst wird durch das sich bei der Diffusion der im Überschuß vorhandenen Säure ausbildende Diffusionspotential auf elektrostatischem Wege hervorgerufen (Abb. 5).

B. Transportvermittler ohne Energiezufuhr. Diese grundsätzlich lehrreichen Modelle können zwar zur Beschreibung der öfter angetroffenen Austauschvorgänge unter den Kationen herangezogen werden, für die Erklärung der Kaliumanreicherung in den Zellen sind sie jedoch nicht verwendbar, weil hier die Konzentration der Kaliumsalze soviel mal höher als die der Säuren ist, daß K^+-

Ionen und nicht die H^+-Ionen potential-bestimmend sind. Außerdem wird bei allen bisher besprochenen Modellen die Membran nur als passives Hindernis für verschiedene Ströme eingesetzt, ohne daß dabei auf die *Vorgänge in der Membranphase* selbst zurückgegriffen wird.

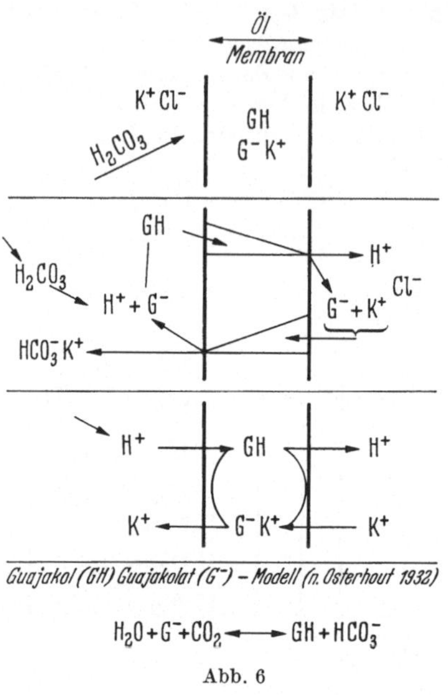

Guajakol (GH) Guajakolat (G⁻) – Modell (n. Osterhout 1932)

$$H_2O + G^- + CO_2 \longleftrightarrow GH + HCO_3^-$$

Abb. 6

Das einfachste Modell, welches auch auf chemische Reaktionen in der 2. Phase Rücksicht nimmt, stammt von OSTERHOUT (1933, 1940)[10] (Abb. 6). Eine ölige Phase trennt 2 wäßrige KCl-Lösungen. Das System wird mit Guajakol gesättigt, welches sowohl frei als auch als K-Guajakolat in beiden Phasen — jedoch im Öl angereichert — vorhanden ist. Einleitung von CO_2 auf einer Seite führt zur Störung der Gleichverteilung, weil zunächst auf dieser Seite aus dem Salz freies Guajakol entsteht; dafür strömt aus der anderen Seite K-Guajakolat nach. Auf diese Weise wird hier K^+ angereichert. Im Öl diffundiert Guajakol von der CO_2-Seite fort und dissoziiert auf der anderen Seite erneut. Das so neu entstandene K-Salz reichert sich im Öl auf dieser Seite an und diffundiert durch die Ölmembran in die CO_2-Seite zurück. Im Effekt entstehen so zwei Ströme; ein Strom der H^+-Ionen aus der CO_2-Lösung und einer der K^+-Ionen in die CO_2-Lösung.

Diese Ströme, getrieben durch die eingeleitete CO_2, werden im Öl deswegen gekuppelt, weil hier beide Formen des Guajakols gut löslich sind. Sie bewegen sich dem Konzentrationsabfall entsprechend im Öl undissoziert in der einen und als Guajakol-Kaliumkomplex in der anderen Richtung. Sie verlassen aber auch die Ölphase. Wäre das nicht der Fall, dann hätte man es hier mit einer Koppelung der beiden Ströme über einen dritten in sich zurückkehrenden, also zirkulierenden Strom zu tun. Solche *zirkulierenden Substanzen*, welche praktisch nur in einer zweiten Phase löslich und am kontinuierlichen Stofftransport beteiligt sind, bezeichnet man als *Carrier*. Man beachte aber, daß grundsätzlich — wie beim Elektroosmose-Modell — die Verknüpfung beider Ströme auch ohne Carrier zu einem aktiven Transport führen kann. Andererseits braucht, wie bald auszuführen sein wird, nicht jede durch einen Carrier vermittelte Stoffbewegung ein aktiver Transport zu sein. Entscheidend ist, daß zunächst überhaupt durch einen Carrier als Lösungsvermittler lipoidunlösliche Stoffe durch lipoide Phasen hindurch befördert werden können.

Der Aufbau der Zellmembran aus *Lipoidmaterial* kann von verschiedenen Gesichtspunkten aus verstanden werden: für die Verknüpfung von Stoffströmen aber ist er deswegen besonders geeignet, weil diese nur in solchen Strukturen stattfinden kann, durch die der ungekuppelte Fluß sehr langsam erfolgt oder ohne Hinzutreten des Carriers gar nicht verlaufen würde. Das trifft allgemein für Lipoidstrukturen zu, weil die zu befördernden Stoffe in der Regel stark hydrophilen Charakter tragen: Eine Pumpe kann nur dort wirksam werden, wo kein in seiner Größe in Betracht kommendes Leck besteht, wo also geeignete Sperrmauern in Form der Lipoidphasen vorliegen. Die Überbrückung kann durch hydrophile eingebaute Abschnitte oder durch bewegliche im Film gelöste Fähren, „Aufnahme-Lipoide", Carrier genannt, im absperrenden „Schutzlipoid"[11], erfolgen.

Ein Carrier ist zunächst ein *beweglicher Transportvermittler*, dessen Bewegung durch die Diffusion in der zweiten Phase, d. h. durch die Wärmeenergie unterhalten wird, ohne daß ein weiteres endergonisches Geschehen im Sinne eines aktiven Transportes vorläge. Da sich der transportierte Stoff in einfachem Zahlenverhältnis, also stöchiometrisch mit der Carriermolekel vereinigt und außerdem nur eine gegebene Menge von letzterem in der Membran vorhanden ist, wird mit steigender Stoffkonzentration sehr bald eine *Sättigung des Trägers* mit dem zu transportierenden Stoff erreicht. Zuvor wird sich je nach der Konzentration ein bestimmtes Verhältnis zwischen freien und gebundenen Plätzen am Träger wie bei einer Adsorption einstellen: die Kurve der Geschwindigkeit des Transportes über einen Carrier trägt den Charakter einer Adsorptions-, also einer Sättigungskurve. Diese Tatsache wurde für manche Stoffe, speziell Farbstoffe, schon vor über 50 Jahren beobachtet und in dem beschriebenen Sinn zuerst 1923 gedeutet (NETTER 1923[11]; E-Kinetik nach WILBRANDT und ROSENBERG 1954[12]). Wie stark ausgeprägt dieser Charakter ist, hängt von der Komplexaffinität ab; ist sie sehr gering, die Komplexdissoziation also groß, dann wird die

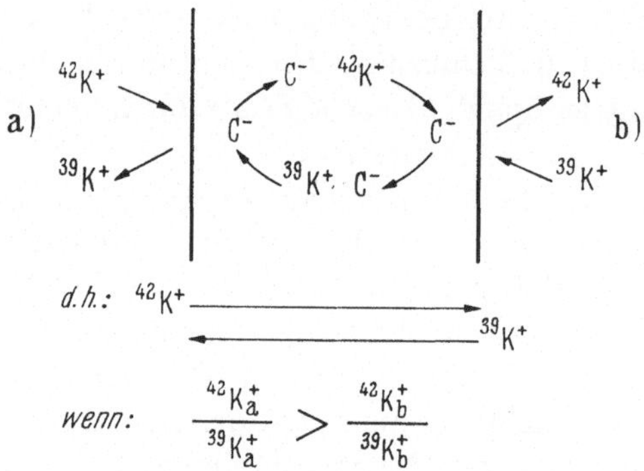

Abb. 7. *Carrier exchange diffusion.* Bei sehr kleinem $^{42}K/^{39}K$-Verhältnis auf einer Seite findet eine Isotopenbewegung nur zu dieser Seite hin statt

Kurve geradlinig, d. h. die Durchschleusungsgeschwindigkeit ist der Differenz der Konzentration beiderseits proportional (D-Kinetik). Trotzdem ist auch in diesem Fall der Träger noch für das Zustandekommen des Übertritts verantwortlich.

Die einfachste durch einen Carrier vermittelte Transporterscheinung konnte erst in Experimenten mit Tracern erkannt werden. Dabei fiel schon früh auf, daß z. B. ^{42}K relativ schnell getauscht wird, obwohl kein Netto-Transport stattfindet und die Membran wie z. B. bei den Erythrocyten als praktisch undurchlässig für Kalium angesehen werden mußte. Tatsächlich können einzelne Kalium-Ionen ganz überwiegend durch Carrier-Transport hin- und herbewegt werden, ohne daß Konzentrationsänderungen der K-Salze beiderseits stattfinden: die anwesenden K-Isotope konkurrieren auf beiden Seiten um die anionische Ladung des Carriers, der nur in Salzform auf die andere Membranseite gelangt. Über die Bewegungsrichtung entscheidet der Unterschied des Isotopenverhältnisses auf beiden Seiten (Abb. 7)[7]. Hier liegt der Typ der *Carrier-exchange-Diffusion* nach USSING[13] (1952) vor.

Dabei kann der Carrier im Gegensatz zum Guajakol beim Osterhout-Modell nur als Salz die Membran in beiden Richtungen passieren; bei diesem „Drehtürmechanismus" müssen beide diagnonalen Fächer besetzt sein. Das braucht für den Transport von Nichtleitern nicht zuzutreffen, hier können die Fähren gewissermaßen auch leer zurückfahren (vgl. Abb. 8).

Nach diesem Schema kann die Carrier-gebundene Aufnahme der Zucker in die menschlichen Erythrocyten oder in die Darmwandzellen bei der Resorption hinein ihre Erklärung finden

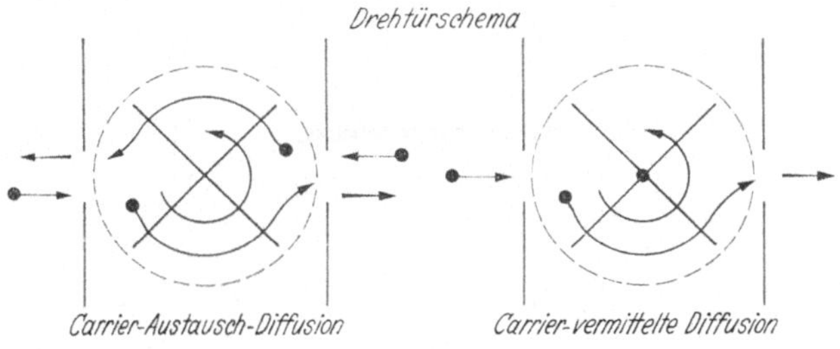

Abb. 8

(Le Fèvre[14] u. a.). In beiden genannten Fällen findet keine oder mindestens nicht immer eine chemische Veränderung am Carrier statt, welche sich unter Aufwendung nicht zu vernachlässigender chemischer Energien vollzöge. Es erfolgt damit auch keine Kuppelung von Stoffströmen unter Energieübertragung über den Carrier. Beide Mechanismen ermöglichen somit auch keinen aktiven Transport, der letztere aber wohl eine Carrier-vermittelte *erleichterte Diffusion* (Danielli 1956)[15]. Sie vollzieht sich an einer Membran, welche ohne den Vermittler keine Diffusion gestatten würde. Es läßt sich darüber hinaus ein Diffusionsvorgang denken, bei dem die Penetration durch geeignete fixierte hydrophile Einlagerungen in die lipoiden Membranen zustandekommt. Sie könnte auf dem Wege über den Tausch von H-Brücken zwischen dem Substrat einerseits und dem H_2O und dem eingelagerten Porenmaterial andererseits vollzogen werden. Formal würde hierbei die zur Diffusion erforderliche Aktivierungsenergie erniedrigt, so daß auch der Ausdruck *katalysierte Diffusion* berechtigt ist.

Weiter kann eine Diffusion so über einen Carrier gekuppelt werden, daß ein Bergauftransport eines Stoffes im Gegenstrom durch das Absinken eines zweiten, in entgegengesetzter Richtung diffundierenden zustande kommt. Dafür ist es nur erforderlich, daß der Carrier — wie bei der Exchange-Diffusion — zu beiden Stoffen eine vergleichbare Affinität besitzt. Er wird z. B. mit dem Stoff A beladen und durch das Konzentrationsgefälle dieser Verbindung in der Membran in eine Richtung getrieben; auf der anderen Seite wird nun A gegen B getauscht und der Carrier dann in der Verbindung mit B durch das nunmehr entgegengesetzt gerichtete Konzentrationsgefälle dieser Verbindung in der Richtung des Stoffes A zurückbewegt. ROSENBERG und WILBRANDT (1957)[16] verifizierten diesen *induzierten Gegentransport* an Erythrocyten mit dem Mannose-Glucose-Paar.

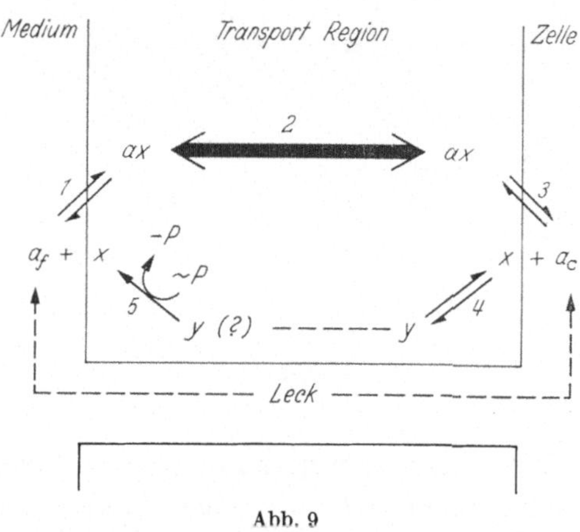

Abb. 9

Schema für den Aminosäuretransport nach HEINZ

C. Transportvermittler mit Energiezufuhr. Es ist sehr wahrscheinlich, daß für einen solchen Leerlauf, bei dem nur osmotische Energie benutzt wird, oft der gleiche Carrier zum Einsatz kommt, welcher bei Übernahme von Energie aus chemischen Reaktionen operiert. Im letzten Fall aber handelt es sich bereits um einen aktiven Transport. Ein Beispiel dafür bietet das Schema von HEINZ[17] (Abb. 9) für die Aufnahme von Aminosäuren in Ascites-Tumorzellen (vgl. auch JOHNSTONE u. QUASTEL, CRANE 1961)[17]. Dabei kann eine schon vorhandene oder erst durch einen passiven Carrier ermöglichte Diffusion mit Hilfe einer chemischen Reaktion durch lokale Erhöhung des Konzentrationsgefälles in der Membran gefördert oder gelegentlich auch erst ermöglicht werden. In diesem Falle findet eine *aktive Diffusionssteigerung* statt. Sie unterscheidet sich im Mechanismus nicht vom aktiven Transport, es sei denn, daß man — wie es häufig geschieht — diesen Terminus allein für die leichter feststellbare Beförderung gegen den Gradienten des chemischen oder elektrochemischen Potentials reserviert (vgl. Tab. 1).

Tabelle 1. *Bewegung gelöster Stoffe*
(ohne Einfluß mechanischer oder elektrischer Felder)

Katalysierte Diffusion	Passive Diffusion (über H-Brücken oder Adsorption)
erleichterte Diffusion exchange diffusion induzierter Gegentransport	durch Carrier *ohne* Aufnahme chemischer Energie
gesteigerte Diffusion „bergauf" Transport	durch Carrier *mit* Aufnahme chemischer Energie

Ein *aktiv transportierender Carrier* erfährt bei jedem Transportschritt eine revertierbare chemische Verwandlung. Die Energie zu dieser Verwandlung wird durch den aktiven Energie- oder Stoffstrom geliefert; er unterhält den chemischen Kreisprozeß am Carrier, der seinerseits wiederum den „bergauf"-Transport der Substanz ermöglicht. Ein Carrier kann jeder in der Membran frei vorhandene Stoff oder jeder hier vorliegende bewegliche Molekülabschnitt sein, welcher durch eine chemische Reaktion derart verändert wird, daß eine Vereinigung mit dem Transportstoff erfolgen kann, infolge derer dieser — sei es durch Diffusion oder durch Lageveränderung in dem Membrangefüge — auf die andere Membranseite gelangt. An dieser wird er durch eine weitere chemische Reaktion unter Rückbildung und Rückdrehung oder Rückdiffusion des freien Carriers in dessen Ausgangszustand und Ausgangslage zurückversetzt. Ein Carrier kann also auch ein festgelegter *Membrananteil mit beweglichen und reaktionsfähigen Gruppen* sein; als solcher kann er wie Lipoide vom Phosphatid-Typ oder wie Proteine oder Peptide polaren Charakter tragen, der durch eine chemische Reaktion, ein elektrisches Feld oder durch die Stoffbindung in seiner Stärke beeinflußbar ist. Im allgemeinen jedoch versteht man unter einem Carrier einen diffusiblen, in der Membran löslichen, frei beweglichen *Stoff in mindestens zwei verschiedenen Zuständen.* Die chemische Natur eines biologisch wirksamen Carriers ist leider noch in keinem Fall gesichert, obwohl im Prinzip jeder lipoidlösliche Stoff mit veränderbaren polaren Gruppen dann ein Carrier sein könnte, wenn er zu Komplexbindungen mit dem Transportstoff befähigt ist.

Eine kurze *energetische Überlegung* (HEINZ und PATLAK[18]) zeigt, daß die beim aktiven Transport aufzuwendende Energie abhängt 1. von der Differenz der elektrochemischen Potentiale des Stoffes, d. h. angenähert dem log. des Quotienten seiner Konzentrationen beiderseits, 2. von dem log. des Quotienten der entgegengesetzt gerichteten Flüsse des bewegten Stoffes und 3. von dem Energiewert der Summe aller der Vereinigung mit dem Carrier vorgeschalteten und der sich nach Freigabe anschließenden Rückreaktionen am Carrier. Diese 3. der drei zu addierenden Portionen ist als irreversibler Energieverlust anzusehen und außerdem und — damit zusammenhängend — auch in ihrer Größe experimentell schwer oder gar nicht zu erfassen, während die beiden anderen Glieder unter günstigen Bedingungen experimentell zugänglich sind.

Für den Reaktionsablauf bei der Energieübertragung werden hauptsächlich zwei Möglichkeiten diskutiert.

a) Die Veränderung des Carriers durch Oxydoreduktionsvorgänge, d. h. es würde die Energie einer Oxydationsreaktion übertragen.

b) Es würden energiereiche Gruppen, z. B. Phosphatgruppen, mit mehr oder weniger geringem Energieverlust übertragen werden.

1. Einfacher Redox-Carrier. Ein Modell eines Ionenkonzentrierungsvorganges mit Hilfe von Redoxenergie wurde zuerst von DAVIES[19] (1948, 1950) zur Erklärung der Säurebildung im Magen entwickelt. In diesem speziellen Fall treibt ein Redoxpotential — im Modell das eines Fe^{2+}/Fe^{3+}-Systems — eine Gaskette, bestehend aus 2 Wasserstoffelektroden, von denen die an den negativen Pol des Redox-Systems geschaltete mit CO_2 durchströmt wird. In dem stromliefernden Redox-Element wird links Fe^{3+} unter Elektronenaufnahme an der Elektrode reduziert, rechts Fe^{2+} durch Elektronenabgabe oxydiert (Abb. 10). Der vom Pt-Blech der H_2-Elektrode rechts gebildete Wasserstoff wird links durch Elektronenabgabe wieder in H-Ionen verwandelt. Hierdurch findet die eigent-

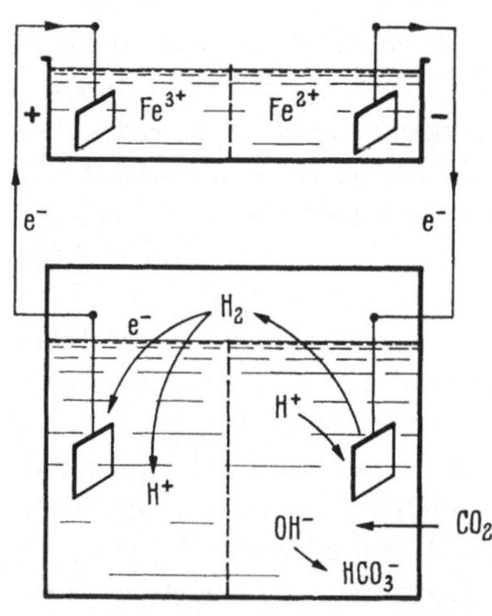

Abb. 10. H^+-Produktion in einem Elektronenkreisvorgang

liche Säurebildung statt. Die rechts nach Fortfangen der H^+-Ionen übriggebliebenen OH^--Ionen werden durch Einleitung von CO_2 unter Bildung von HCO_3^-- weggefangen. Auf diese Weise wird durch das energieliefernde Redox-System eine wesentlich höhere H-Ionen-Konzentration erreicht, als durch alleinige Zugabe der CO_2 geschaffen werden kann.

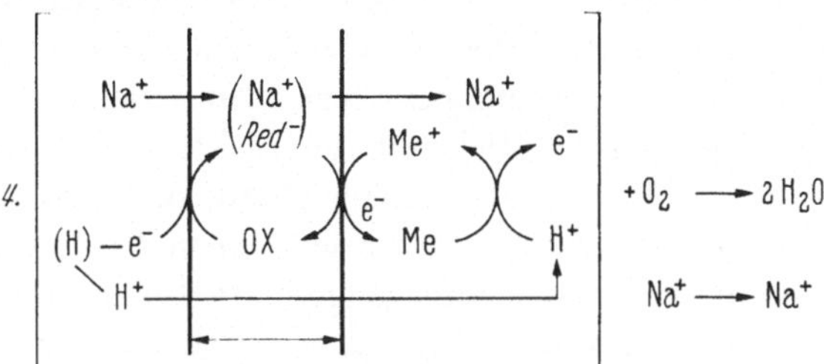

Abb. 11a. Redox-Carrier-Na+-Pumpe

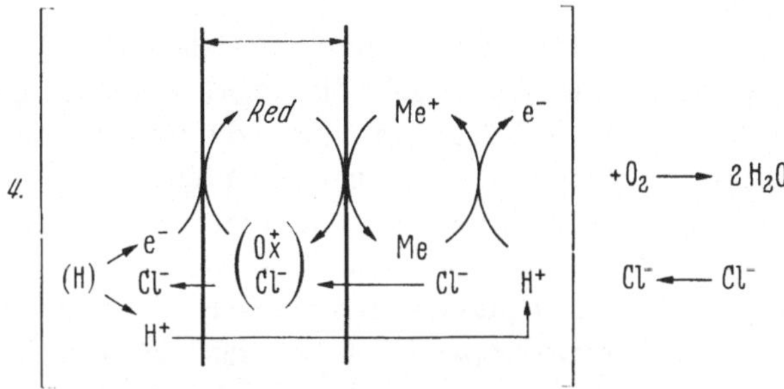

Abb. 11b. Anionen-Redoxpumpe

Die Schwierigkeit der Übertragung dieses Modells liegt in der Frage der Existenz einer Elektronenleitung in vivo. Die Möglichkeit dieses Vorganges im Eiweiß ist oft diskutiert worden. Der Wasserstofftransport könnte über die Atmungskette erfolgen, ebenso wie die ständige Elektronenlieferung aus dem Redoxvorgang durch Oxydationsprozesse unterhalten werden kann. Im ganzen gesehen ist das Modell aber zu allgemein gehalten, um auf die chemischen Verwandlungen eines Carriers direkt angewandt werden zu können. Außerdem fordert das angegebene Schema keineswegs direkt eine Carrier-Substanz; es sei denn, daß der vielleicht an Flavine gebundene Wasserstoff als verkappter Carrier für H^+-Ionen betrachtet wird.

Conway[20] (1951/1955) hat die Redox-Hypothese auf den Transport anderer Kationen angewandt, indem er die spezielle Vorstellung von einem spezifisch auf K^+-, Na^+-Ionen usw. eingestellten Membran-Carrier entwickelte (Abb. 11a u. 11b). Es ist für diese Modelle erforderlich, daß auf beiden Seiten der Membran ein verschiedenes Redoxpotential herrscht, so daß ein Elektronenstrom von der negativen reduzierenden zur positiven oxydierenden Seite fließt. Die Kuppelung findet wieder über den bisher unbekannten lipoidlöslichen Carrier statt, welcher entweder im oxydierten Zustand als Kation oder ungeladen, oder im reduzierten Zustand ungeladen oder als Anion vorliegt. Die anionische Form dient dem Transport der Kationen, die kationische dem der Anionen. Dementsprechend erfolgt die Bewegung der Kationen in der Richtung, die der Anionen entgegen der Richtung des Elektronenstromes durch die Membran. Eine Kombination beider Pumpen, der Anionen- und der Kationen-Redoxpumpe oder die mit einem Typ passiv selektiver Permeabilität für die entgegengesetzte Ionenart kann eine aktive Salzbewegung verständlich machen oder die Bildung der Salzsäure erklären, bei der sowohl die H^+- wie die Cl^--Ionen aktiv bewegt werden.

Zur Idee der Redox-Kationenpumpe, welche auch für die H^+-Ionenkonzentrierung im Magen sehr diskutiert wird, gelangte CONWAY[20] auf Grund seiner Erfahrungen an gärender und atmender Hefe. Sie tauscht große Mengen an H^+-Ionen gegen zugesetzte K^+-Ionen der Außenflüssigkeit aus und kann auf diese Weise der Außenlösung einen p_H-Wert von 1,4—2,0 erteilen, also in Cl^--Gegenwart HCl erzeugen. Es hat sich gezeigt, daß dieser Kationentausch nicht passiv nach dem Gesetz der partiellen Gleichgewichte an kationenpermeablen Membranen — wie er oben erörtert wurde — erfolgt. Er kann nämlich durch geeignete Konzentrationen von Azid oder DNP zum Erliegen kommen, ohne daß dabei die Gärung der Hefe merklich beeinflußt wird. Die Hefe scheidet auch wie der Muskel im Austausch gegen K^+- aktiv Na^+-Ionen aus. Dieser Austausch wird durch Blausäure und unter anaeroben Bedingungen gehemmt. Die Hemmung durch Anoxie findet jedoch nicht bei Gegenwart von 5% Glucose statt. Unter diesen Bedingungen läßt sich eine Beeinflussung des Tauschvorganges durch zugesetzte organische und auch anorganische Redox-Systeme feststellen.

Dabei zeigte sich in den Versuchen von CONWAY[20], daß offenbar eine *kritische Schwelle des Redoxpotentials* außen besteht, unterhalb derer der Tausch nicht zustande kommt. Mit steigendem E_h-Wert des in geringer Menge zugesetzten Systems wächst die Geschwindigkeit des aktiven Tausches mit dem Redox-Carrier. CONWAY[20] vermutet, daß die oxydierte Form des Carriers mit dem reduzierten Farbstoff zu einer stabilen Verbindung zusammentritt, welche den Redoxcyclus in der Membran blockiert.

Ob umgekehrt die Förderung durch elektropositivere Systeme auf einer Steigerung der Geschwindigkeit der Elektronenaufnahme durch den Carrier beruht, ist unsicher.

Nach CONWAY[21] existieren mehrere Kationencarrier in der Hefemembran, von denen ein allgemeiner, in der Hauptsache mit K^+-Ionen arbeitender und ein spezieller mehr im Sauren um p_H 4,5 wirkender gesichert zu sein scheinen: letzterer ist weitgehend spezifisch für den Transport von Mg, welcher sich praktisch nur unter aeroben Bedingungen und in Gegenwart von Glucose vollzieht[21] (CONWAY und BEARY 1958[21], ROTHSTEIN 1954[23], 1960[21]; NETTER und SACHS 1961[21]). Der spezielle Carrier für Mg^{2+} ist nur wenig gegenüber K^+-Ionen empfindlich. Für den Na^+-Ionentransport an der Muskelmembran hat CONWAY[21] weiter gezeigt, daß eine kritische Barriere des Membranpotentials besteht, unterhalb derer die aktive Bewegung von Na^+ nach außen nicht zustande kommt. Im ganzen ergeben sich viele Hinweise auf die Existenz eines transportierenden Redoxsystems, ohne daß jedoch sein Vorkommen bis heute als bewiesen angesehen werden kann.

2. *Phosphatgekuppelter Redox-Carrier.* Eine energetische Überlegung ergibt weiter, daß der Redox-Carrier-Vorgang mit einem *hohen Nutzeffekt* arbeitet; z. B. ist der Extrasauerstoffverbrauch in der Magenschleimhaut so groß, daß für eine O_2-Molekel 12 H^+-Ionen entgegen einer Konzentration von $n/10$ gebildet werden, während vorstehende Modelle nur die Bildung von 4 H^+- oder die Beförderung von 4 Na^+- oder K^+-Ionen gestatten. Ähnliches gilt für die Na^+-Beförderung an der isolierten Froschhaut oder bei pflanzlichen Zellmembranen. Um diesen hohen Nutzeffekt, der sich vor allem auch bei der Bewegung bei fehlendem oder geringem Konzentrationsgefälle zeigt, zu erklären, wurde zuerst von OGSTON[26], später auch von CONWAY[20] ein *zusätzlicher energetischer Beitrag* aus der Quelle energiereicher Phosphate mit herangezogen. Mit ihrer Hilfe wird letzten Endes eine bedeutende Steigerung der Konzentration des zu transportierenden Stoffes auf einer Seite der Membran erreicht. Der zu konzentrierende Stoff ist hier eigentlich das Elektron, bzw. die reduzierte Form des Carriers. Die möglichen Konzentrierungseffekte ergeben sich aus dem Wert der Spaltungsenergie der eingesetzten Phosphatverbindungen. So läßt sich unter der Annahme einer schnellen, d. h. fast gleichmäßigen Verteilung der durch ATP phosphorylierten reduzierten Form des Trägers, dessen Konzentration für die freie Form auf beiden Seiten durch die unterschiedlichen Bildungs- und Spaltungsgeschwindigkeiten stark verschieden gehalten wird, das maximal erreichbare Verhältnis

eben beider Konzentrationen aus der Gleichgewichtskonstanten der Gesamtreaktion entnehmen:

$$(CarH_2)_i + ATP \rightarrow (CarH_2)_a + P + ADP.$$

Unter Einsetzen von 11 kcal für die Gesamtreaktion, erhält man so eine 10^8fach höhere Konzentration von $(CarH_2)_a$, d. h. auf der Seite, an der die Spaltung der Carrierphosphatverbindung vollzogen wird (Abb. 12). Diese Überlegung zeigt, daß und wie die Anwendung von ATP-Energie zur gerichteten Konzentrationssteigerung gestaltet werden kann. Sie zeigt weiter, daß die Richtung eines mit ihrer Hilfe aufgebauten Konzentrationsgefälles von der unterschiedlichen Anwesenheit oder Verankerung verschieden wirksamer Enzyme auf beiden Membranseiten abhängt.

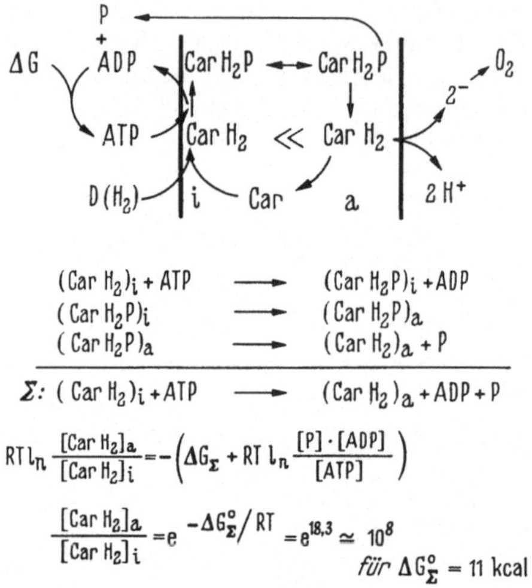

$$\begin{array}{lll} (Car\,H_2)_i + ATP & \longrightarrow & (Car\,H_2P)_i + ADP \\ (Car\,H_2P)_i & \longrightarrow & (Car\,H_2P)_a \\ (Car\,H_2P)_a & \longrightarrow & (Car\,H_2)_a + P \\ \hline \Sigma:\ (Car\,H_2)_i + ATP & \longrightarrow & (Car\,H_2)_a + ADP + P \end{array}$$

$$RT\ln\frac{[Car\,H_2]_a}{[Car\,H_2]_i} = -\left(\Delta G_\Sigma + RT\ln\frac{[P]\cdot[ADP]}{[ATP]}\right)$$

$$\frac{[Car\,H_2]_a}{[Car\,H_2]_i} = e^{-\Delta G^0_\Sigma/RT} = e^{18,3} \simeq 10^8 \quad \textit{für } \Delta G^0_\Sigma = 11\ \text{kcal}$$

Abb. 12. ATP-getriebener Redox-Carrier

Daß *enzymatische Wirkungen an Strukturen* und namentlich auch an membranartige Strukturen fixiert sind und durch hier lokalisierte Enzyme zustande kommen, ist seit langem bekannt — man braucht in diesem Zusammenhang nur an den Fermentbestand und die Membranstruktur der Mitochondrien zu denken. Auch für das Vorkommen von Phosphatasen an den Zellmembranen existieren zahlreiche Hinweise einschließlich histochemischer Befunde (z. B. MÖLBERT[23]). Für die Gegenwart von ATPasen in den Membranen und Hüllen markloser Tintenfischnerven gab LIBET[23] schon 1948 experimentelle Beweise; das gleiche wurde für die Membranen der Erythrocyten und für ihre Schatten bewiesen. Für die Außenmembranen von Bakterien wurde es vor kurzer Zeit von ABRAMS[24] (1960) wiederum einwandfrei belegt. Es wäre zuviel Skepsis, wenn man nicht versuchen würde, die an die Membranen zu lokalisierenden Reaktionen mit mehr oder weniger energiereichen Phosphaten in die Vorstellungen vom Stofftransport durch die Grenzschichten einzubauen.

Der oben beschriebene Weg einer Konzentrationssteigerung geeigneter Redoxzustände eines Carriers mit Hilfe von Phosphaten ist nicht die einzige Form einer Energieübernahme. Grundsätzlich kann die Energieübertragung an jeder Stelle im Kreisprozeß der Carrierzustände erfolgen und es ist wahrscheinlich, daß mehr *Zwischenstufen* in diesem Geschehen existieren, als bisher erörtert zu werden brauchten. Es wäre zweifellos denkbar, daß phosphorylierte Zwischenprodukte ein geändertes, z. B. negativeres Redoxpotential erzeugen und damit erst Reduktionen ermöglichen oder in einem Ausmaß veranlassen würden, wie sie ohne diese Zwischenschaltung nicht zustande kommen könnten. Man wird hierbei an das analog gelagerte Problem der Reversion der oxydativen Phosphorylierung, d. h. der Reduktion mit Hilfe von ATP, erinnert, für deren Existenz KLINGENBERG und BÜCHER[22] in letzter Zeit gute Belege geliefert haben (1961). In beiden Fällen müssen jene energiereichen Zwischenglieder berücksichtigt werden, von denen aus bei normalem Lauf der Elektronen die Phosphatübertragung auf ADP erfolgt. Man muß annehmen, daß die sog. *ATPase* der Membran in allen diesen Fällen phosphorylierend, d. h. *als Transphosphorylase* wirkt. Daß ein geschwinder Umsatz von Phosphaten mit Membranbestandteilen vor sich geht, läßt sich den Versuchen HILLMANNS[25] entnehmen. Er studierte mit Hilfe von ^{32}P die Aufnahme anorganischen Phosphates in Erythrocyten, welche relativ langsam erfolgt. Dabei ließ sich aber feststellen, daß eine schnelle Einlagerung von Phosphaten in die Membranen oder die Blutschatten stattfindet. Die chemische Natur dieses interessanten *primären Membranphosphats* ist nicht bekannt; es ist vor allem noch nicht klar, ob das für die erleichterte Diffusion des Phosphats verantwortliche Membranphosphat Beziehungen zu Phosphatidsäuren besitzt. Offenbar ist jedoch mit diesen Versuchen eine erhebliche Anreicherung einer Zwischenverbindung des zu transportierenden Stoffes nachgewiesen.

Auch die Modellvorstellungen müssen eine größere Zahl solcher Verbindungen einbeziehen. Dabei soll hier abgesehen werden von den *verschiedenen Zuständen eines Na-transportierenden Carriers,* welche zur Erklärung feinerer elektrophysiologischer Äußerungen herangezogen wurden und bei denen ein ruhender und aktiver Normalzustand von einem inaktiven unterschieden wird (vgl. LÜTTGAU[27]). Schon die einfache Tatsache, daß Erythrocyten salzundurchlässig sind — und es zur Aufrechterhaltung ihrer osmotischen Struktur auch sein müssen — im Verein mit der hohen Permeabilität für Anionen verbietet eine Kationenpermeabilität; sie läßt aber einen langsameren Gegentausch von K^+ gegen Na^+ zu. Er findet tatsächlich unter Aufrechterhaltung der Bilanz beiderseits im gegenseitigen Tausch so statt, daß die Fluxe für beide je in beiden Richtungen gleich sind. Da die Bewegung beider Kationen carriergebunden ist, stellt man sich im Anschluß an SOLOMON nach SHAW vor, daß die *Carrier für beide* unter Zuführung von ATP-Energie *ineinander verwandelbar* sind[27] (SOLOMON 1952, SHAW 1954). Die der Abb. 13 zugrundeliegende Vorstellung rechnet nicht explicite mit einem Redox-Carrier, aber sie rechnet damit, daß nur mehr oder weniger undissoziierte Alkalimetallkomplexe die Membran durchdringen können, während die geladenen Carriersubstanzen an der Grenzschicht verbleiben und hier verwandelt werden. Energetisch ist das Schema möglich, wenn die Übernahme der Energie beim Übergang von X^- zu Y^- langsamer verläuft als der entgegengesetzte durch Enzyme der anderen Seite katalysierte Vorgang, denn nur dann kommt der

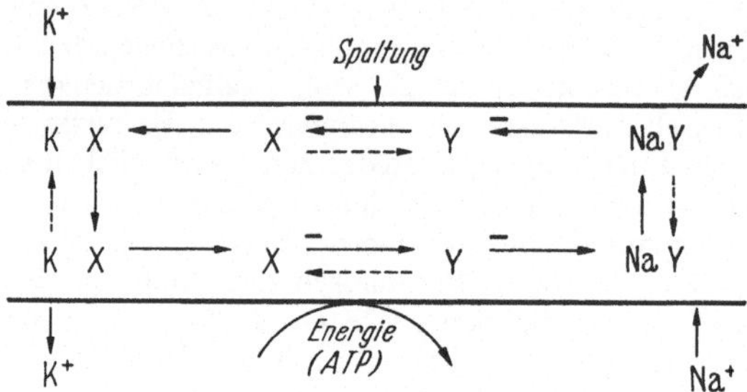

Abb. 13. Gekuppelter Na^+- und K^+-Redoxcarrier

mit dem Transport und der Carrierverwandlung verknüpfte Energiefluß in der angegebenen Richtung zustande. Das Schema gibt im übrigen von der Tatsache Rechenschaft, daß ein aktiver Na-Transport von innen nach außen nicht ohne Gegenwart von K^+-Ionen außen zustandekommt. Andererseits ist der Transport der K^+-Ionen nur bis zur Erreichung eines Carrier-Sättigungswertes von der äußeren K^+-Ionen-Konzentration abhängig.

3. Phosphatgekuppelte Lösungsvermittler. Es ist bemerkenswert, daß unter den Nucleosidtriphosphaten die Substanz ATP zur Unterhaltung des aktiven Transportes am wirksamsten ist, obwohl die biochemischen Reaktionen im übrigen, wenn es sich um die Lenkung der Synthese bestimmter Stoffgruppen handelt, auch UTP, GTP oder CTP verwenden. Für die reine Energielieferung und Phosphorylierung wird jedoch auch im Stoffwechsel hauptsächlich von ATP Gebrauch gemacht. Danach läßt sich vermuten, daß auch beim aktiven Transport nicht mehr als diese beiden Funktionen für die primären Leistungen des ATP in Frage kommen. Außer der Energielieferung hat die *Phosphorylierung* hier nun sicher auch die Aufgabe, den *lipoidlöslichen Carrier zur Komplexbildung mit Ionen* zu befähigen. Man weiß nur noch nicht, wie die lipoidlöslichen Komplexe mit den Kationen beschaffen sind und kann sich nur allgemein vorstellen, daß Paraffinketten mit anionischen Resten die

Metallionen binden, wobei die polaren Teile in der Mitte liegen und durch die apolaren Reste nach außen abgeschirmt werden. Unter dem Einfluß des elektrischen Feldes einer Doppelschicht zwischen Wasser und Öl würden die Komplexe deformiert und die wasserlöslichen Ionen unter Hydratation in die wäßrige Phase gebracht. Für den umgekehrten Vorgang ist eine geeignete Lage des Carriers die Voraussetzung zur Schaffung des Komplexes, d. h. sie wird die erforderliche Aktivierungsenergie für seine Bildung weitgehend herabsetzen können.

Im Rahmen dieser allgemeinen Überlegungen hat ein spezieller und wie es scheint sehr charakteristischer Befund Bedeutung erlangt: der aktive *Ionentransport*, in erster Linie der Na-Transport aus den Zellen heraus, speziell zunächst aus den Erythrocyten, wird *durch Herzglucoside spezifisch gehemmt.* Diese Entdeckung von SCHATZMANN[28] (1953) ist vielfach bestätigt und näher untersucht worden. Die Wirksamkeit hängt von der Konfiguration des Glucosides an C_{17} des Steroidringes und von dem Vorliegen eines ungesättigten Lactonringes an dieser Stelle ab. Der sterische Antipode (β) ist ebenso wie alle untersuchten Steroidhormone praktisch unwirksam. Die Wirkung erstreckt sich nicht auf das ATP-liefernde System, sondern auf den Transportmechanismus und mit Sicherheit, wenn nicht allein auf das transphosphorylierende Enzym; daher wird an Erythrocyten auch der Eintransport von K^+-Ionen empfindlich gestört. Die Wirksamkeit der Cardiaca ist auch hier sehr hoch. Unter der Voraussetzung, daß je ein solches Molekül mit je einem Transportort in der Zellmembran reagiert und hier verankert bleibt, errechnet GLYNN (1958) aus den minimal wirksamen Konzentrationen, daß nur 1000 *Orte für den aktiven Transport* auf der gesamten Oberfläche eines Erythrocyten existieren. Da dessen Oberfläche rd. $1 \cdot 10^{10}$ Å^2 beträgt, würde unter Annahme eines wohl kaum zu großen Areals von 1000 Å^2 pro Transportort damit (d. h. mit 10^6 Å^2) nur ein hundertstel Prozent der Gesamtoberfläche durch die für den aktiven Transport zugänglichen Stellen besetzt sein, eine Überlegung, welche die experimentell vor allem in chemischer Beziehung bestehenden Schwierigkeiten verdeutlicht.

Die auffallende Spezifität der Strophanthinwirkung veranlaßte KELLER[29] zur Prüfung der Frage, ob der Carrier selbst ein Steroid sei, so daß die Hemmwirkung als Konkurrenz beider um den Enzymort aufzufassen wäre. Er bemühte sich, diese Frage in Modellversuchen zu klären, bei denen als Carrier der *3-Phosphorsäureester des Cholesterins* diente. Er wurde in Olivenöl gelöst, mit welchem eine für Wasser dichtschließende Filtrierpapiermembran getränkt war. Wie manche Phosphatidsalze, speziell des Cephalins, besitzen die Alkalisalze des Cholesterylphosphates eine beträchtliche Öllöslichkeit. Zusatz von radioaktiven ^{86}Rb-Salzen führte dementsprechend zum leicht meßbaren Übertritt auf die andere Seite beim Vorliegen von im übrigen gleichkonzentrierten Rb-Salzen beiderseits. Hinzufügen einer wirksamen Phosphatase auf der Gegenseite bewirkt hier eine Spaltung des Cholesterinesters und eine Abgabe des Phosphates als Alkalisalz an die wäßrige Lösung. Hierbei wurde in niedrigen Salzkonzentrationen auch ein Bergauftransport des Rb festgestellt. Der aktive Transport wird in diesem Fall von dem Vorrat an Estern in der Membran gespeist, welche in vivo unter Verbrauch von ATP aufgebaut würden. Aber in Gang gesetzt werden kann der aktive Prozeß erst durch die Esterspaltung auf der Gegenseite. Es handelt sich hier um den ersten *Modellfall* für den aktiven Transport, welcher *unter Verwendung von Enzymen* realisiert wurde (Abb. 14). Dabei hat sich aber leider ergeben, daß nur ungereinigte Nierenphosphatase in der Lage war, den Cholesterinester zu spalten. Auch Versuche mit Cholesterinsulfat waren erfolgreich. Aber für das Vorkommen von Cholesterin-(SO_4)-Estern spricht bisher kein biochemischer Befund, außer der Tatsache, daß mehrere Sulfatasen existieren, welche diese Ester gut spalten.

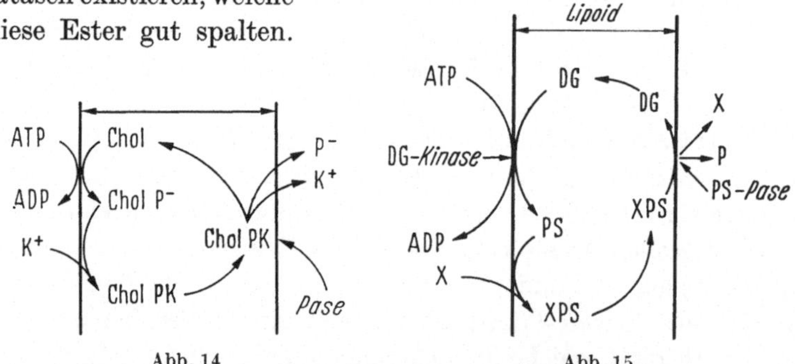

Abb. 14. Cholesterinester-Modell nach KELLER (1959)

Abb. 15. *Phosphatidsäure-Carrier* (HOKIN u. HOKIN, 1959). DG = Diglycerid; PS = Phosphatidsäure; PS-Pase → PS-Phosphatase

Dieses Modell verwirklicht also die Verknüpfung zweier Flüsse, z. B. dem des Phosphates und des Rb über einen zirkulierenden Carrier. Damit bietet es eine gute Analogie zu dem vor 2 Jahren von HOKIN und HOKIN[29] aufgestellten und inzwischen mehrfach diskutierten Schema, welches die enzymatische Bildung von *Phosphatidsäuren* aus Diglyceriden zur Grundlage nimmt, also einen Vorgang, der sich bei der Synthese von Phosphatiden abspielt (Abb. 15). Das von den Autoren vorgeschlagene Schema ist zur allgemeinen Anwendung geeignet, da die Natur des Stoffes insoweit offengelassen wird, als nur von hydrophilem Material gesprochen wird. Es könnte auch für den Transport von Ionen in Frage kommen und speziell auch für den von Phosphaten selbst, z. B. auch aus energiereichen Verbindungen. Es hat den Vorteil, daß das Vorkommen von — durch Acetylcholin weiter aktivierbaren — Diglyceridkinasen für Mikrosomen aus Gehirn und Drüsen sichergestellt ist. Vor allem aber hat das Ehepaar HOKIN[29] gezeigt, daß ^{32}P aus ATP besonders schnell in die Phosphatidsäuren des Gehirns eingebaut wird, und ganz neuerdings ist von japanischen Autoren[30] (YOSHIDA und NUKADA 1961) berichtet worden, daß auch der Phosphatid-turnover in Gehirnschnitten durch K^+-Ionenzugabe ausgesprochen gesteigert wird. Nach HOFFMAN[36] wird auch das ^{32}P aus ATP besonders schnell in die Phosphatidsäure der Erythrocytenlipoide eingebaut. Da die Lipoide der roten Blutkörperchen sich fast ausschließlich in den Membranen befinden, kann dieser Befund als wichtig, jedoch keineswegs als allein beweisend angesehen werden. Es kommt aber hinzu, daß Strophanthin auch den Einbau in die Phosphatidsäure ebenso verhindert, wie es den aktiven Transport lähmt.

Diese eben gebrachten Schemata sind untereinander recht ähnlich und in dieser Form wohl zuerst von DRABKIN[31] (1948) für den Zuckertransport bei der Resorption gegeben worden. Das einfache Phosphorylierungsschema mußte hier jedoch aufgegeben werden; ebenfalls wird die Bildung von nichtionischen Metaphosphorsäureestern der Zucker[32] (TH. ROSENBERG) nicht mehr diskutiert. Über den Mechanismus der Wirkung von *Permeasen* ist heute ebensowenig auszusagen wie über die Bedeutung von soeben im Rattenzwerchfell gefundenen *Glucan-Peptiden*[33] (O. WALAAS 1960; G. KESSLER und W. J. NICKERSON). Ob die zuletzt genannten Zwischenstoffe überhaupt als Carrier anzusehen sind, ist durchaus noch nicht zu übersehen, ebensowenig, wie auszuschließen ist, daß nicht überhaupt häufiger, als bisher angenommen, eine *Kuppelung* von Flüssen *ohne Carrier* erfolgen kann.

Bei der Erörterung dieser Frage sollen die von SCHLÖGL[35] entwickelten Vorstellungen, welche mit einem Sortiereffekt zwischen den Ionen bei ihrer Mitnahme durch einenin verschieden weiten Poren entstehenden elektroosmotischen Kreisstrom rechnen, beiseite gelassen werden. Man kann auch, wie weitere Modellversuche von KELLER[34] zeigten, zwei Ströme ohne zirkulierende Carriersubstanz nur auf Grund geeigneter Löslichkeiten einzelner Glieder in einer öligen Phase miteinander zu einem aktiven Transportgeschehen verbinden. Sein Katecholsulfatmodell beruht darauf, daß aus z. B. 4-Nitro-Brenzkatechin und ATP über aktives Sulfat Brenzkatechinsulfat aufgebaut wird. Diese Substanz ist auch nicht öllöslich, aber sie nimmt das nicht öllösliche Methylenblau als öllösliches Komplexsalz mit *Katecholsulfat* in das Öl hinein. Auf der anderen Seite wird durch eine hier vorhandene Arylsulfatase das Sulfat vom Katechol abgespalten. Der Methylenblau-Komplex zerfällt hierbei und die Methylenblaukationen werden trotz des schon vorhandenen Farbstoffes auf dieser Seite angereichert. Gleichzeitig geht auch Brenzkatechin hier in die Lösung (Abb. 16). Es ist nur ein durch wechselnde Löslichkeit im Öl bedingter gradueller Unterschied, wenn praktisch nicht von einem zirkulierenden Carrier gesprochen wird. Dieser Versuch gibt ein weiteres Modell, bei dem auf enzymatischem Wege chemische Energie in Konzentrationsarbeit verwandelt wird.

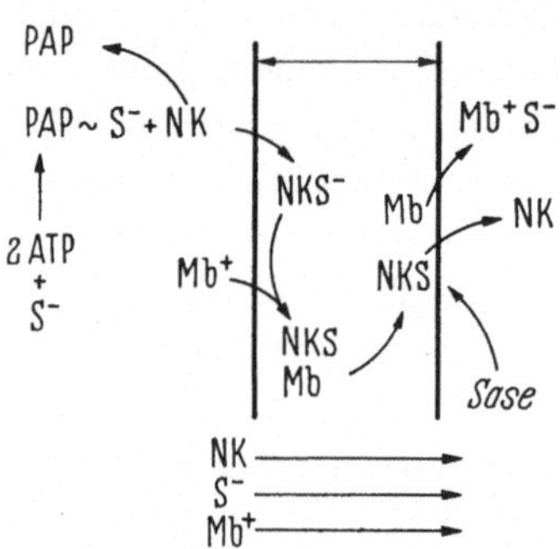

Abb. 16. *Katechol-Sulfat-Modell* nach KELLER (1960). S^- = Sulfat; PAPS = aktives Sulfat; NK = 4-Nitro-Brenzkatechin; Mb^+ = Methylenblaukation; Sase = Sulfatase

4. *Transportleistung und Aktivität der Membrantransphosphorylasen.* Biologisch wichtig ist eine richtige *Einstellung des Ausmaßes der Pumpwirkung.* Sie wird wie in jedem Regelkreis von der Regelgröße, hier also von der Konzentration des zu transportierenden Stoffes, abhängen. Die Mittel, über die dieser Regelkreis arbeiten kann, müssen einmal der in seiner Größe veränderliche Widerstand der Membran für die entsprechende Substanz und andererseits die Aktivität der am Transport beteiligten Enzyme sein. Darunter ist auch die *langsame Adaptation* an ein neues Substrat und damit die

Induktion der Enzymbildung zu verstehen, wie sie z. B. in der Bereitstellung der entsprechenden Permeasen zum Ausdruck kommt. Insbesondere aber ist die *schnelle Anpassung* an akut auftretende Konzentrationsänderungen der einzustellenden Stoffe bzw. Ionen zu beachten. Hierzu haben in letzter Zeit besonders Skou, Hoffman und Post[36] wesentliche experimentelle Beiträge geliefert. Sie betreffen das phosphat- und energieübertragende Enzym der Zellmembranen, welches hier kurz ATPase genannt wurde und welches nach den meisten gegebenen Schemata das Transportgeschehen einleitet.

Skou[36] untersuchte die ATPase der Membranen markloser Nerven, die beiden anderen Autoren die der Erythrocyten. Die Ergebnisse aller 3 Arbeitskreise über die Abhängigkeit der Fermentaktivität von der Konzentration jener Ionen, die an diesen Membranen aktiv bewegt werden, decken sich vollständig. In beiden Geweben sind praktisch die einzig in Betracht kommenden Strukturen in unserem Sinne *aktive Membranen.* Geeignete Zentrifugate der entsprechenden Homogenate zeigen nun sehr charakteristische *Abhängigkeiten* ihrer Aktivität *vom Ionenmilieu* unter der Voraussetzung, daß Mg-Ionen in physiologischen Konzentrationen zugegen sind. Sie lassen sich alle unter dem Gesichtspunkt des Schemas von Shaw über die Verknüpfung des K^+- und Na^+-Transportes an Erythrocyten-Membranen dann verstehen, wenn die Geschwindigkeit des Transportes allein durch die Änderung der Enzymaktivität über die geänderte Konzentration von Na- und K-Ionen, d. h. über die Regelabweichung, bestimmt wird. Die Versuche lassen sich so deuten, daß entweder 1 Enzym mit 2 Kationenbindungsorten oder 2 Enzyme mit je einem Ort existieren. Die Orte unterscheiden sich in ihrer Affinität zu beiden Ionen insofern, als der vermutlich nach außen gelegene schon bei physiologischer, d. h. relativ niedriger K^+-Konzentration, aber relativ hoher Na^+-Konzentration nahezu mit K^+ abgesättigt ist, während der andere Ort im Gegensatz zu dem ersten bei recht niedriger Na^+-Konzentration — dem Zellinnern entsprechend — für Na und bei hoher für K abgesättigt ist. Das Ausmaß der Absättigung wird aus der Aktivität geschlossen; denn sie wird durch beide Ionen entsprechend der soeben geschilderten Abhängigkeit gesteigert. Nach den Vorstellungen von Skou[36] wirken die Alkalikationen dadurch, daß sie eine Chelatverbindung mit einem Mg-ATP-Enzymkomplex eingehen. Dabei hat es sich ergeben, daß 2 besonders aktive Konzentrationszonen existieren. Dieses Verhalten spricht einerseits für das Vorkommen der ATPasen auf beiden Membranseiten und zweitens dafür, daß dieses System tatsächlich mit dem Verhalten der Ionen etwas zu tun hat. Daß es sich aber hierbei eben um die aktive Bewegung der Ionen selber handelt, geht nun daraus hervor, daß *Strophanthin*, welches den aktiven Ionentransport hemmt, gerade nur für diese ATPase ein sehr stark wirksames Enzymgift ist. Es ist *bei der ATPase* aus Erythrocyten *in gleicher Konzentration wie bei der Lähmung des aktiven Transportes*, nämlich bei etwa 10^{-7} m, *wirksam* (POST 1960[36]). Diese Befunde geben den hier referierten Vorstellungen einen besonders hohen Grad von Wahrscheinlichkeit. Wenn man nun noch hinzunimmt, daß Strophanthin, wie schon kurz berichtet, auch den Einbau von ^{32}P in die Phosphatidsäuren des Blutkörperchenstromes hemmt (HOFFMAN 1960[36]), wird auch die Hypthese von HOKIN[29] gestützt.

III. Ausblick

Im ganzen betrachtet unterbaut der Komplex der soeben referierten Ergebnisse über das Verhalten der ATPase mindestens ebenso gut wie viele elektrophysiologische Einzelerfahrungen die eingangs bevorzugte Auffassung von der Bedeutung der Membranstruktur für den aktiven Transport und für die passiven cellularen Verteilungsvorgänge. Es ist zu erwarten, daß die Weiterentwicklung unseres Wissens auf diesem Gebiet, die in chemischer Hinsicht besonders dringend erforderlich ist, von den begonnenen enzymchemischen Untersuchungen ihren Ausgangspunkt nehmen wird. Wie weit jedoch die hier geäußerten Vorstellungen von der Betätigungsweise eines Membrancarriers wirklich Bestand haben, bleibt abzuwarten. Ich habe die Hoffnung, daß diejenigen, welche dann den Stein der Weisen besitzen, nicht über unsere Bemühungen so reden müssen, wie es PETRARCA einst über die Alchimisten tat: Rauch, Asche, Schweiß und leere Worte (fummo, ceneri, sudori ed parole).

Literatur

1 BEST, J. B., and J. Z. HEARON: In Mineral Metabolism Bd. I, 11. New York and London; ed. by Comar and Bronner 1960; KEDEM, O.: Criteria of Active Transport. Symp. Membr. Transport. Prag 1960, S. 51, u. A. KATCHALSKY, ebenda S. 58.; NETTER, H.: Theoretische Biochemie. Berlin-Göttingen-Heidelberg: Springer-Verlag 1959.

2 Koefoed-Johnsen, V., and H. H. Ussing: Acta physiol. scand. **28**, 60 (1953).
3 Hartley, G. S.: Trans. Faraday Soc. **33**, 1021 (1937).
4 Kuhn, W., u. K. Ryffel: Hoppe Seylers Z. physiol. Chem. **276**, 145 (1942).
5 Hargitay, B., u. W. Kuhn: Z. Elektrochemie **55**, 539 (1951); W. Kuhn: Klin. Wschr. **37**, 997 (1960).
6 Ullrich, K. J.: Ergebn. Physiol. **50**, 433 (1960); Schimassek, H., D. Kohl u. Th. Bücher: Bioch. Z. **331**, 87 (1959).
7 Netter, H.: Pflügers Arch. ges. Physiol. **220**, 107 (1928).
8 Kostyuk, P. G., and Z. A. Sorokina: Symp. Membr. Transp. and Metab. Czechoslow. Acad. Sci. S. 93. Prag 1960.
9 Teorell, T.: J. gen. Physiol. **21**, 107 (1937).
10 Osterhout, W. J. V., and W. M. Stanley: J. gen. Physiol. **15**, 667 (1932); Osterhout, W. J. V.: Cold Spr. Harb. Symp. quant. Biol. **8**, 51 (1940).
11 Netter, H.: Pflügers Arch. ges. Physiol. **198**, 225 (1923).
12 Wilbrandt, W., u. Th. Rosenberg: Helv. physiol. pharmacol. Acta. **9**, C 86 (1951); W. Wilbrandt: In Moderne Probleme der Pädiatrie IV, 30. Basel 1959.
13 Ussing, H. H.: Advanc. Enzymol. **13**, 21 (1952).
14 Le Fèvre, P. G.: Symp. Soc. exp. Biol. **8**, 118 (1954).
15 Danielli, J. F.: Disc. Faraday Soc. Nr. 21, 238 (1956).
16 Rosenberg, Th., u. W. Wilbrandt: J. Gen. Physiol. **41**, 289 (1957).
17 Heinz, E., u. P. M. Walsh: J. biol. Chem. **233**, 1488 (1959); Johnstone, R. M., and J. H. Quastel: Biochim. biophys. Acta **46**, 527 (1961); Crane, R. K.: Biochim. biophys. Acta **46**, 477 (1961).
18 Heinz, E., and Cl. S. Patlak: Biochim. biophys. Acta **44**, 324, 1960; vergl. Cl. S. Patlak: Bull. Math. Biophys. **18**, 271 (1956); **19**, 209 (1957).
19 Davies, R. E., and A. G. Ogston: Biochem. J. **46**, 324,1950; vergl. E. E. Crane, R. E. Davies and N. M. Longmuir: Biochem. J. **43**, 321 (1948).
20 Conway, E. J.: Science **113**, 270 (1951). Biochemistry of gastric acid secretion. Springfield, Ill.: Cl. Thomas 1953; Int. Rev. Cytol. **4**, 377 (1955); Conway, E. J., T. G. Brady and E. Carton: Biochem. J. **47**, 369 (1950); Conway, E. J., and R. Kernan: Biochem. J. **61**, 32 (1955).
21 Conway, E. J., and M. E. Beary: Biochem. J. **69**, 275 (1958); Rothstein, A.: Symp. Membr. Transp. and Metab. Czechoslov. Acad. Sci. S. 173. Prag 1960; Jennings, D. H., D. C. Hooper and A. Rothstein: J. gen. Physiol. **41**, 1019 (1958); Netter, H., u. L. Sachs: Bioch. Z. **334**, 18 (1961); Conway, E. J.: Regul. Inorganic ion cont. of cells, Ciba Symp. S. 2, London 1960.
22 Klingenberg, M., u. Th. Bücher: Bioch. Z. **334**, 1 (1961).
23 Mölbert, E., Fr. Duspiva u. O. v. Deimling: Histochemie **2**, 5 (1960); Libet, B.: Fed. Proc. **7**, 72 (1948); Rothstein, A.: Protoplasmatologia Bd. II E4. Wien 1954.
24 Abrams, A., P. Mc. Namara and F. B. Johnson: J. biol. Chem. **235**, 3659 (1960).
25 Schauer, R., u. G. Hillmann, Hoppe-Seylers Z. physiol. Chem. **325**, 9 (1961).

[26] DAVIES, R. E., A. G. OGSTON (19), E. HEINZ and K. ÖBRINK: Physiol. Rev. **34**, 643 (1954).
[27] SHAW, T. J.: Ph. D. Thesis. Cambridge University 1954; GLYNN, I. M.: Progr. Biophys. Biophys. Chem. **8**, 241 (1957); SOLOMON, A. K.: J. gen. Physiol. **36**, 57 (1952); LÜTTGAU, M. C.: Experientia (Basel) **12**, 482 (1956).
[28] SCHATZMANN, H. J.: Helv. physiol. pharmacol. Acta **11**, 346 (1953).
[29] KELLER, H.: Ber. ges. Physiol. **215**, 44 (1960); HOKIN, L. E., and M. R. HOKIN: J. biol. Chem. **234**, 1381, 1387 (1959). Prager Symposion S. 103.
[30] JOSHIDA, H., and T. NUKADA: Biochim. biophys. Acta **46**, 408 (1961).
[31] DRABKIN, D. L.: Proc. Amer. Diab. Ass. **8**, 171 (1948).
[32] ROSENBERG, TH.: Rep. Steno Hosp. (Kbh.) **4**, 59 (1950).
[33] WALAAS, O., B. BORREBAEK, T. KRISTIANSEN and E. WALAAS: Biochim. biophys. Acta **40**, 562 (1959); KESSLER, G., and W. J. NICKERSON: J. biol. Chem. **234**, 2281 (1959).
[34] KELLER, H.: Angew. Chem. **21**, 778 (1960).
[35] SCHLÖGL, R.: Zum Materietransport durch Porenmembranen. Habli.-Schrift. Göttingen 1957.
[36] SKOU, J. C.: Biochim. biophys. Acta **42**, 6 (1960); Prager Symp. S. 124; POST, R. L., C. R. MERRITT, C. R. KINSOLVING and C. D. ALBRIGHT: J. biol. Chem. **235**, 1796 (1960); Prag. Symp. S. 115; HOFFMAN, J.: Regul. inorg. ion cont. of cells London 1960, Ciba Symp. S. 92.

Diskussion

Diskussionsleiter: HEINZ, *Frankfurt/M.*

Wir danken Herrn Prof. NETTER für seinen reichhaltigen eleganten Vortrag, durch den die verschiedenen Austausch- und Kupplungsvorgänge in den Hirnzellen erheblich in Bewegung gesetzt worden sind. Es fragt sich nun, sollen wir gleich fortfahren mit der Diskussionsbemerkung von Herrn KLEINZELLER?

Herr KLEINZELLER hat die Freundlichkeit gehabt, trotz Visumschwierigkeiten von Prag hierher zu kommen, um hier zu dem Vortrag von Herrn Prof. NETTER eine Diskussionsbemerkung zu machen.

KLEINZELLER (Prag): Das von USSING 1958 vorgeschlagene Modell[1] des Mechanismus des gekoppelten aktiven Na^+ und K^+ Transportes in Epithelzellen (Froschhaut, Nierenrindenzellen) beruht u. a. auf den folgenden Tatsachen:

1. Einer polaren Struktur der Epithelzelle, die RHODIN[2] elektronenmikroskopisch für die Tubularzelle erwiesen hat; der aktive Kationentransport wäre dann nur an einer Membranwand lokali-

[1] USSING, H. H.: IVth. Intern. Congr. Biochem., Symp. No. III, Preprint No. 12. Wien, 1958.
[2] RHODIN, J.: Publ. Dept. Anatomy, Karolinska Institutet, Stockholm 1955.

siert, und zwar unserem Vorschlag gemäß[3] an der basalen Zellmembran. Dieser Vorstellung entsprechen auch die Potentialmessungen von GIEBISCH[4].

2. Die Froschhaut kann aktiv Na^+ nur in Anwesenheit von K^+ transportieren. Da nun der Transport von Na^+ und K^+ in vielen Zellen nicht äquivalent verläuft (in der Nierenrinde finden wir ein Verhältnis von etwa 2 transportierten Na^+ per akkumuliertes K^+), setzt das Ussingsche Schema voraus, daß die Membran, unabhängig von dem Kationenträger-Mechanismus außerdem auch frei durchgängig für K^+ ist.

3. Jedes Modell des aktiven Transportes sollte auch die direkte Energiequelle definieren und die Tatsache erklären, daß in Epithelzellen allgemein ein Na/O_2 Quotient von mehr als 12 (meistens nun etwa 18) gefunden wird.

Wir haben nun einige der erwähnten Aspekte in Nierenrindenschnitten näher untersucht und ich möchte kurz die Resultate erwähnen.

Die erste Frage lautet, ob wirklich die Nierenrindenzellen vollkommen frei für K^+ durchgängig sind. Alle, die sich mit den Eigenschaften der Nierenrindenschnitte befaßten, finden, daß auch bei längerem Aufbewahren der Schnitte in isotonischem NaCl bei 0° (oder in Anwesenheit von Inhibitoren, die den aktiven Transport blockieren), nur ein Teil des Gewebs-K^+ ausgelaugt wird. Wir haben nun eingehend den Charakter dieser K^+-Fraktion untersucht[5]. Wie aus Tab. 1 ersichtlich ist, finden wir mittels drei verschiedener Methoden praktisch denselben Wert der K^+-Fraktion, die bei 0° nicht auslaugbar ist, die in 2 Std bei 0° oder 25° (aerob) nicht mit ^{42}K austauschbar ist, und die sich nicht an dem Donnan Gleichgewicht bei 0° und 25° beteiligt. Diese Resultate sind in guter Übereinstimmung mit denen von FOULKES[6] für die Austauschbarkeit von K^+ durch Rb^+. Auch ein Teil des Gewebs-Na^+ scheint nicht frei beweglich zu sein. Daraus folgt, daß in diesem Gewebe die Flux-analyse von Kationen, besonders aber von K^+, sowohl als die Analyse der Elektrolytendistribution den „gebundenen" Anteil

---

[3] CORT, J. H., u. A. KLEINZELLER: J. Physiol. **133**, 287 (1956).

[4] GIEBISCH, G.: Circulation **21**, 879 (1960).

[5] KLEINZELLER, A., A. KNOTKOVÁ u. K. JANÁČEK: I. Sjezd čsl. biofysikální Společnosti, Abstr. Nr. 42 (1961).

[6] FOULKES, E. C.: Fed. Proc. im Druck (1961).

Tabelle 1. *„Gebundene" K^+-Fraktion in Nieren-Rindenschnitten des Kaninchens.* Resultate in m-Equiv. K/100 g Trockengew.

K nicht auslaugbar in 0,154 M NaCl bei 0°	11,2 ± 0,7	m-Equiv.
K nicht austauschbar mit ^{42}K in 2 Std bei 0° und 25° .	14,7	m-Equiv.
K nicht an Donnan Gleichgewicht beteiligt (aus der Donnan Verteilung des Cl berechnet) . .	11,5	m-Equiv.

Tabelle 2. *Geschwindigkeitskonstanten, Donnan Quotienten für Cl und K, im stationären Zustand der Nieren-Rindenschnitte.* Resultate ± s. E.

Temperatur	0°		25°	
$[K^+]_0$	5,6	103	5,6	103
k_1 (Influx) $\times$ Min^{-1}	0,52 ±0,05	0,177±0,006	2,09±0,08	0,30±0,013
k_2 (Efflux), $\times$ Min^{-1}	0,34± 0,04	0,245±0,005	0,195±0,01	0,27±0,012
k_1/k_2	1,53	0,722	10,7	1,11
$[Cl^-]_0/[Cl^-]_i$	1,495±0,037	1,10	2,49	1,29
$[K^+]_i/[K^+]_0$ (unkorr.)	8,9 ±0,3	1,13	17,0	1,47
$[K^+]_i/[K^+]_0$ (korr.)	1	0,83	8,8	1,26

in Betracht ziehen muß; so könnte man sich erklären, warum GIEBISCH[4] niedrigere Potentialwerte fand als der Donnan-Verteilung des K^+ entspricht.

Wir haben weiter beide ^{42}K-Fluxe im stationären Zustand in Nieren-Rindenschnitten gemessen[5] (siehe z. B. Abb. 1). Werte der Geschwindigkeitskonstanten der ^{42}K-Fluxe (exchange constant von USSING[7]) nach Korrektion für den extracellulären Raum, und auf intracelluläres Wasser bezogen, bei 0° und 25° sowie die Donnan Verteilungsquotienten von K^+ und Cl^- sind in Tab. 2 angegeben.

Aus der Tabelle ersieht man, daß 1. die Konstanten für den Efflux derselben Ordnung sind, unabhängig von der K^+-Konzentration im Medium und der Temperatur. 2. Die Influx-Konstante ist nur bei 25° und $[K^+]_0 = 5{,}6$ mM erhöht, wo ein intensiver aktiver Transport von Kationen (und Wasser) zustande kommt; bei hohem $[K^+]_0$ (103 mM) scheint daher der aktive Transport blockiert zu sein, wie wir schon durch chemische Analyse früher fanden[8]. Es steht offen, durch welchen der verschiedenen möglichen Mecha-

[7] USSING, H. H.: Advanc. Enzymol. **13**, 21 (1952).

[8] KLEINZELLER, A.: Membrane Transport and Metabolism — A Symposium, S. 527. Publ. House Czechosl. Acad. Sci., Prague 1961.

nismen hohes $[K^+]_0$ den aktiven Transport blockiert. 3. Bei 0° ist das Verhältnis der Geschwindigkeitskonstanten in experimentellen Fehlergrenzen mit dem Donnan-Verhältnis des Cl^- und auch des K^+ identisch, soweit das letztere für die „gebundene" K^+-Fraktion korrigiert wird. Bei 25° wurden größere Unterschiede nur bei 5,6 mM $[K^+]_0$ gefunden; dies ist im Einklang mit der Funktion der Kationenpumpe. 4. Vorläufige Resultate zeigen, daß die Geschwindigkeitskonstanten für ^{24}Na bei 0° zwar ordnungsgemäß denen des ^{42}K entsprechen; allerdings ist weder die Na-Distribution, noch das Verhältnis der Fluxe in Übereinstimmung mit den entsprechenden Werten für K^+ und Cl^-. Es ist nicht ausgeschlossen, daß das Karrier-System auch bei 0° operativ sein könnte.

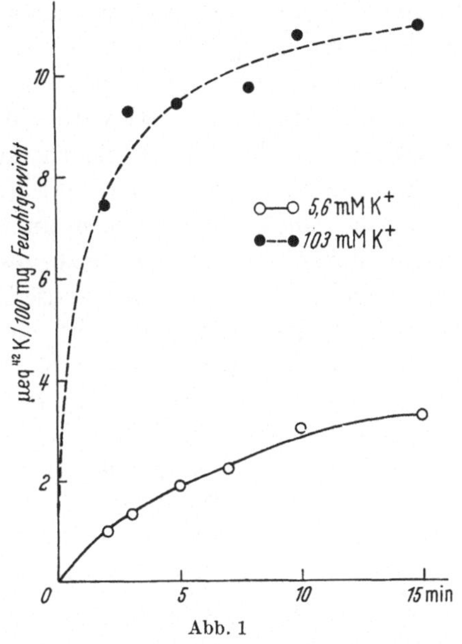

Abb. 1

Unter verschiedenen Bedingungen finden wir, daß aus den Tubularzellen Wasser als eine praktisch isotonische Lösung von Elektrolyten transportiert wird[8]; ob nun Wasser durch separate Membranporen, unabhängig von dem Kationen-Karriersystem sich bewegt, steht unseres Erachtens nach offen.

Aktiver Kationentransport in Nierenrindenschnitten kann auch anaerob, und zwar in neugeborenen Tieren erwiesen werden[9]; das glykolytische System liefert hier die notwendige Energie, da dieser Transport durch mM Jodacetat vollkommen blockiert wird.

Die erwähnten Resultate scheinen also grundsätzlich im Einklang mit dem Ussing Modell zu sein, bzw. mit demjenigen von Davies und Keynes[10].

---

[9] Čapek, K. u. Kleinzeller: Čsl. Fysiol. **9**, 6 (1960).

[10] Davies, R. E., and R. D. Keynes: Membrane Transport and Metabolism — A Symposium, S. 336. Publ. House Czechosl. Acad. Sci., Prague 1961.

In Anbetracht wiederholter Behauptungen (z. B.[11]) sei hier betont, daß das Acetylcholinesterase System nicht an dem aktiven Kationentransport in der Nierenrinde oder Gehirnrinde teilnimmt; selektive Inhibitoren dieses Systems hemmen nämlich diesen Transport nicht[12].

Ich möchte noch die oft erwähnte Funktion der -SH Gruppen in dem Mechanismus des Kationentransports berühren. Wir betrachten diese Frage als offen. Sehr niedrige Konzentrationen von Quecksilberverbindungen, oder auch N-Ethyl-Maleinimid und Formamidin-Disulphid (etwa 10^{-5} M) erhöhen das Froschhaut-Potential als Folge einer Erniedrigung der Chlorid-Permeabilität; bei höheren Konzentrationen sinkt das Potential und die Na^+ und Cl^--Permeabilität steigen enorm an [13]. In Nierenrindenschnitten, wo eine starke Bindung von Hg zustandekommt (bis 10 mMol/100 g Trockengew.) wurde keine markante Beeinträchtigung des aktiven Transportes gefunden[14], eher aber eine bedeutende Erhöhung der passiven Membranpermeabilität für Elektrolyte und auch Harnstoff. Eine Rolle der -SH-Gruppen in den Membranporen kann also nicht ausgeschlossen werden.

Die Frage wird oft gestellt, ob und wieweit eine Korrelation zwischen in vitro Messungen des Kationentransports und der physiologischen Funktion berechtigt ist. Wir haben nun die Na-Reabsorption in der Niere des Hundes in vivo mit dem Maximal-Transport von Na^+ in Nieren-Rindenschnitten eines Hundes desselben Gewichts verglichen. Die Menge reabsorbierten Na^+ kann aus dem Volumen des Glomerularfiltrats und ausgeschiedenem Na auf etwa 14 Equiv./Tag berechnet werden; die maximale Geschwindigkeit des Na^+-Transports in vitro entspricht einem aktiven Transport von etwa 29 Equiv./Tag in der Nierenrinde. Eine vorsichtige Korrelation zwischen in vitro und in vivo Versuchen scheint also berechtigt zu sein.

HEINZ: Ich danke Herrn Prof. KLEINZELLER für seinen interessanten Beitrag, ich glaube wir haben jetzt genügend Stoff zum Diskutieren. Ich bitte um Wortmeldung.

---

[11] KIRSCHNER, L. B.: Nature (Lond.) **172**, 348 (1953).

[12] KLEINZELLER, A., u. J. H. CORT: Coll. Czechosl. Chem. Comm. **23**, 746 (1958).

[13] JANÁČEK, K.: I. Sjezd. čsl. biofysikální společnosti, Abstr. No 35 (1961).

[14] KLEINZELLER, A., u. J. H. CORT: Physiol. bohemoslov., **10**, 349 (1961).

[15] CORT, J. H., u. A. KLEINZELLER: J. Physiol. **142**, 208 (1958).

HASSELBACH (Heidelberg): Ich möchte anknüpfen an die Vorstellungen von HOKIN u. HOKIN. Wenn ich mich recht erinnere, ist dieses Modell zur Klärung des Transportes von einwertigen Kationen entwickelt worden. Wenn das Modell aber richtig ist, d. h. wenn also zweiwertiges Phosphat beteiligt ist, sollte es sich auch eignen zum Transport von zweiwertigen Kationen. Ich glaube, wir haben so etwas ähnliches im Muskel gefunden. Es existieren im Muskel kleine Bläschen, die wir ja beschrieben haben, die also einen Durchmesser von 500—1000 Å haben und deren Membrandicke vielleicht 50 Å beträgt. Nun ist es möglich, zu zeigen, daß diese Bläschen aktiv Calcium transportieren und zwar kommt es — wenn man die Bedingungen nur geeignet wählt — zu enormen Anreicherungen von Calcium. Dabei ist vielleicht noch 1. etwas beizutragen zur Frage der Energetik: Diese Calciumspeicherung ist nur möglich, wenn ATP gespalten wird; und zwar gilt also sehr genau, daß für ein Calcium ein ATP umgesetzt wird. 2. ist, glaube ich etwas beizutragen zur Frage der Regulation. Diese Pumpe läuft nur so lange Calcium in der Außenlösung vorhanden ist. Sie stellt ihre Tätigkeit ein, wenn das Calcium verschwunden ist. Das sieht dann etwa so aus: man beginnt mit einer sehr kleinen ATP-aseaktivität, d. h. die Ruheaktivität ist sehr viel niedriger als die, die von Herrn USSING für seine Objekte angegeben worden ist. Nun setzt man Calcium zu. Es kommt dann zu einer starken Aktivierung der ATP-ase, die mit Sättigung der Ca-Akkumulation aufhört. Wenn man nun das Schicksal des Calciums verfolgt, sieht man, daß es eben während der Zeit der ATP-aseaktivität aufgenommen wird, und zwar weitgehend mit der Geschwindigkeit, mit der das ATP gespalten wird.

HEINZ: Ich danke Ihnen, Herr HASSELBACH. Ich muß noch etwas nachtragen: ich habe eben erfahren, daß gerade über dieses Thema Herr PASSOW gleich sehr ausführlich berichten wird. Und darum möchte ich vorschlagen, daß wir die Diskussion, die die Phosphatidsäuren und ihre Funktion betreffen, noch etwas zurückstellen — falls Sie damit einverstanden sind — und vielleicht dann noch zu diesem Beitrag von Herrn HASSELBACH Stellung nehmen werden.

QUADBECK (Homburg/Saar): Untersuchungen an der Blut-Hirn-Schranke lassen außer dem von H. NETTER angeführten Transportmechanismen noch einen weiteren aktiven Stofftransport, den Glucosetransport aus dem Blut ins ZNS, als weiteren Transport-Typ wahrscheinlich werden. Hier geschieht der Glucoseübergang Insulin-unabhängig über einen Mechanismus, der nach Untersuchungen von A. GEIGER Leber-abhängig ist und Beziehungen zum UDPG-System hat. Auf Grund unserer Befunde erfolgt dieser Transport auf folgende Weise: Glucose und Galaktose werden auf der Blutseite laufend in die im Schrankenbereich nachweisbaren Polysaccharid-Strukturen eingebaut. Auf der Hirnseite erfolgt ein laufender Abbau dieser Strukturen, deren Bruchstücke soweit sie Derivate der Glucose sind, für den oxydativen Stoffwechsel verwandt werden. Bei cerebralen Ernährungsstörungen kommt es zuerst zu einem verstärkten Abbau dieser Strukturen, wobei der relative Anteil an Galactosederivaten im Liquor, wie im Hirngewebe ansteigt. Der Energie-bedürftige Teil dieses Transportes dürfte die Überführung von

Glucose in UDPG bzw. die ATP abhängige Bildung des für diese Umsetzung benötigten UTP sein.

HESS (Heidelberg): Unter den Mechanismen des aktiven Transportes gewinnt die ATP-getriebene Redoxpumpe durch neuere experimentelle Befunde zum Mechanismus der oxydativen Phosphorylierung eine besondere Bedeutung. Arbeiten von KLINGENBERG sowie CHANCE, HOLLUNGER, HAGIHARA und HESS haben gezeigt, daß die oxydative Phosphorylierung reversibel ist. Durch ein hohes ATP/ADP + P_a-Verhältnis gelingt es, Pyridinnucleotid und seine Redoxpartner durch Bernsteinsäure sowie auch durch die reduzierten Cytochrome unter gegebenen Bedingungen zu reduzieren. In der Bilanz handelt es sich dabei um eine reduktive Dephosphorylierung. Wir konnten kürzlich nachweisen, daß die Umkehr der oydativen Phosphorylierung auch in intakten Zellen (Ascites-Tumor-Zellen) vorkommt. Zellphysiologisch hat die Atmungsumkehr eine Bedeutung für das Zellwachstum und möglicherweise für den aktiven Transport von Säuren und Elektrolyten. Eine Lokalisation der Mitochrondrien an der Innenfläche der Zellwandung bringt einen engen Kontakt mit den Elementen des aktiven Transports mit sich, die bei einem hohen ATP/ADP + P_a-Verhältnis durch diesen Mechanismus reduziert gehalten werden können. Darüber hinaus möchte ich noch darauf hinweisen, daß Mitochrondrien eine Membran besitzen und alle Eigenschaften des aktiven Transports aufweisen, die hier eng mit der oxydativen Phosphorylierung zusammenhängen. Elektrolyt- und Wasserbewegungen begleiten die oxydative Phosphorylierung. Untersucht man z. B. die Konzentrationen von Na und K in Mitochondrien, deren Atmungskette auf die verschiedenen Gleichgewichte (nach CHANCE) eingestellt ist, so findet man unterschiedliche K^+/Na^+ Quotienten: Für den Zustand „4" einen Wert von etwa 1,18, für den Zustand „3" einen Wert von etwa 0,21.

NETTER: Wenn ich nur ein paar Worte dazu sagen darf: Ich habe in meinem Manuskript auch auf die Umkehr der oxydativen Phosphorylierung im Zusammenhang mit der möglichen Beeinflussung des Redoxpotentials hingewiesen. Ich finde, es ist auch gut und notwendig, auf die Mitochondrien als Modell für die Zellmembranen überhaupt hinzuweisen; aber ob wirklich entsprechende Stellen oder Bruchstücke der Atmungskette in den äußeren Zellmembranen gelegen sind oder nicht, wie es ja z. B. nach CONWAY zu fordern wäre, das wissen wir heute nicht mit Sicherheit; ich persönlich möchte darin sehr vorsichtig sein und sagen, daß ich heute noch sehr skeptisch bin, ob man in der Lage sein wird, chemisch einen Beweis dafür zu führen, weil die Zahl der aktiven Transportorte in der Membran so sehr gering ist.

HOLZER (Freiburg): Setzt man verarmten Hefezellen unter anaeroben Bedingungen Glucose zu, so nimmt die Orthophosphatkonzentration in den Zellen ab und erreicht nach etwa 30 min ein niedriges stationäres Niveau. Studiert man unter denselben Bedingungen das Eindringen von Orthophosphat aus dem Medium, so sieht man, daß in den ersten 30 min kein oder nur ein minimaler Eintritt in die Hefezellen erfolgt. Dieser beginnt erst, nachdem das stationäre, niedrige Niveau des Orthophosphates in den Hefezellen erreicht ist [HOLZER, Biochem. Z. **324**, 148 (1953)]. Es muß demnach ein Regulationsmechanismus bestehen, der dafür sorgt, daß der Eintritt von Phosphat in Hefezellen nur dann stattfindet, wenn in den Zellen Phosphat-verbrauchende Reaktionen in einem solchen Umfange abgelaufen sind, daß das stationäre Phosphatniveau auf einen tiefen Wert abgesunken ist.

STAUDINGER (Gießen): Das was der Biochemiker „Mikrosomen" nennt, ist morphologisch ein Teil des Ergastoplasmas, das, wie elektronenoptisch gezeigt wird, oft mit der Zelloberfläche in Verbindung steht und somit auch für den Stoffaustausch der Zelle verantwortlich sein kann. Andererseits findet man biochemisch eine Reihe von Elektronentransportmechanismen in den „Mikrosomen", deren funktionelle Bedeutung bislang nicht erklärt werden konnte. Dieser mikrosomale Elektronentransport hat sicher nichts mit der oxydativen Phosphorylierung zu tun. Es ist denkbar, daß die mikrosomalen „Redoxasen" mit Redox-carriern verknüpft sind. Das würde funktionelle und morphologische Befunde an „Ergastoplasma", d. h. Mikrosomen, verknüpfen.

NETTER: Wenn ich darauf gleich antworten darf: die Ergastoplasma-Anteile zeigen tatsächlich z. B. auch Phosphatideinbau, gemessen am P-32-Einbau in Phosphatide; ja, in den Mikrosomen des Gehirns haben HOKIN u. HOKIN zunächst einmal diese Versuche mit dem Einbau des Phosphats durchgeführt. Aber nun muß ja nicht der Einbau des Phosphates in entsprechende Partikelchen unmittelbar mit dem aktiven Transport etwas zu tun haben. Folgender Befund könnte es zwar indirekt wahrscheinlich machen; aber er besitzt wegen des Bestehens auch anderer Möglichkeiten für die Betätigung der Phosphatasen keine direkte Beweiskraft: Man hat bestimmte — wenn auch wie immer mit großer Kritik zu bewertende — histochemische Anhaltspunkte dafür, daß ein großer Teil der Phosphatasen, die wir ja auch für die Carrier-Mechanismen benötigen, nur auf den Außenmembranen vorhanden ist (MÖLBERT). Wie weit das sicher ist, weiß ich nicht; aber ich halte es durchaus für möglich, da z. B. bestimmte Phosphatasen nur auf den sonst ja durchaus ähnlich aussehenden Außenmembranen nachzuweisen sind, während die inneren Membranen des Ergastoplasmas diese Enzyme offenbar nicht tragen. Daß das für die Mikrosomen eigentümliche Oxydationssystem auch an den Transportorten der Membranen wirkt, ist aus dem schon zuvor angedeutetem Grunde heute nicht experimentell beweisbar.

BORST PAUWELS (Nijmegen): In Experimenten mit radioaktivem Phosphat habe ich gefunden, daß das markierte Phosphat zuerst (innerhalb einer Sekunde bei 0° C) im ATP, GTP und UTP eingebaut wurde, also eher als in die Zuckerphosphate in die Nucleosid-di(tri)-phosphate übernommen wird. Es interessiert mich nun sehr, ob die Außenmembran der Hefe frei zugänglich ist für Orthophosphat oder nicht. Im letzten Falle nämlich muß man sich vorstellen, daß das Phosphat an der Außenmembran in die Nucleosidtriphosphate inkorporiert und in Form dieser Substanzen durch die Membran transportiert würde.

NETTER: Ich glaube, daß ich das Wort in bezug auf den Phosphattransport eigentlich an Herrn PASSOW weitergeben kann. Aber zuvor möchte ich vielleicht darauf hinweisen, daß die Hefe ja einen Außenrand besitzt, der nach CONWAY ungefähr 10% vom Volumen der Hefezelle ausmacht. Er ist in vieler Beziehung gut durchlässig. Und dann kommt erst der große Stop, der auch vom Phosphat nicht ohne weiteres überwunden wird. Dieser Hinweis würde dann also zu der Gegenfrage von mir führen: Ist es möglich, daß die Prozesse, von denen Sie sprechen, in dem Vorfeld der Hauptmembran stattfinden, von dem ROTHSTEIN vor allen Dingen gesprochen hat, und der durch eine besondere Membran noch nach außen abgegrenzt sein soll?

HOLZER: Orthophosphat dringt nur sehr langsam in Hefezellen ein. Wahrscheinlich ist die Eintrittsgeschwindigkeit des Orthophosphats ein geschwindigkeitsbegrenzender Vorgang beim Wachstum und bei der Zellvermehrung. CONWAY hat eine „outer metabolic region" beschrieben, die nur einen kleinen äußeren Anteil des Hefevolumens umfaßt. In diese Region dringt z. B. Succinat sehr leicht und schnell ein, während es in das Innere der Hefe nur sehr langsam oder gar nicht eindringt. Es kann sein, daß Phosphat in diesen äußeren Bezirk der Hefe wesentlich rascher eindringt als in die innere Region, in der die Phosphat-abhängigen Synthesen für die Wachstumsprozesse stattfinden.

KLAUS (Mainz): Zur Frage der spezifischen Hemmwirkung der Herzglykoside auf den aktiven Transport. Wir haben Versuche durchgeführt, die sich mit dem Einfluß von Herzglykosiden auf den K-Flux am Herzmuskel befassen. Dabei ergab sich, daß therapeutische Konzentrationen von Digitoxigenin (die nur positiv inotrop wirken) keinen Einfluß auf den K-Austausch haben. Toxische Konzentrationen (Kontraktur!) dagegen steigern den K-Efflux beträchtlich und haben nur eine geringe Hemmwirkung auf den K-Influx.

Wie lassen sich diese Befunde mit der hervorgehobenen spezifischen Beeinflussung des aktiven Transportes vereinbaren, oder welche anderen Erklärungsmöglichkeiten für den K-Transport bestehen unter diesen Bedingungen?

USSING (Kopenhagen): Well, perhaps I could answer this question in relation to what Prof. KLEINZELLER already has said, namely that part of the K^+ exchange of living cells is associated with permeation through pores. It is not assumed to be influenced by inhibition of the pump. So far I do not see any contradiction in the observation. However there are other observations which do not quite agree with the assumption that the heart glycosides are specifically associated with kation transport. I think the first observation was made by COOPERSTEIN*, who found that in the large intestine the excretion of bicarbonate is also inhibited by strophantine and we have found in the frog skin preparation that the diffusion of chloride is definitely reduced and even the diffusion of water is reduced during the action of strophantine. Whether this is a direct effect on the membrane or whether it is brought about by a secondary change in the cell following the stopage of the pump, we cannot say, but offhand it seems as if even the passive diffusion of anions is influenced.

* COOPERSTEIN, J. L., and S. K. BROCKMANN: Clin. Res. **6**, 277 (1958).

Heinz: Ich glaube, im Interesse der Zeit müssen wir jetzt die Diskussion unterbrechen, ich wollte nur noch fragen, ob Herr Kleinzeller noch ein Schlußwort wünscht. Ich glaube, einige Fragen waren auch an Sie gerichtet.

Kleinzeller: Ich möchte nur betonen, daß ich mit Herrn Prof. Holzer übereinstimme über die Funktion des Phosphats. Wir haben gewisse Resultate, die in Übereinstimmung sind, auch was den Einfluß von Kalium auf den Phosphataustausch in der Hefezelle betrifft, und 2. noch zu der Frage über die gerade der Kollege aus Mainz sprach. Ich glaube, es ist auch eine Frage, ob und wie schnell die Herzglykoside in die Zelle eindringen, um dort einen Einfluß zu haben. Ich glaube, das wird auch in Betracht gezogen werden müssen bei niedrigen Konzentrationen der Herzglykoside.

Heinz: Ich danke den Herren Vortragenden und Diskussionsrednern und darf ankündigen, daß bis 12 Uhr eine Pause ist.

Zusammenwirken von Membranstruktur und Zellstoffwechsel bei der Regulierung der Ionenpermeabilität roter Blutkörperchen

Von

Hermann Passow

Physiologisches Institut der Universität Hamburg

Mit 15 Textabbildungen

Im vorliegenden Referat möchte ich anhand einer Auswahl experimenteller Untersuchungen über die Ionenpermeabilität roter Blutkörperchen einige aktuelle Probleme der Permeabilitätsforschung skizzieren. Um den Überblick über ein umfangreiches Material zu erleichtern, möchte ich zunächst dogmatisch darstellen, wie Membranstruktur und Zellstoffwechsel bei der Regulierung der Ionenverteilung zwischen Zellinnerem und Plasma zusammenwirken und erst anschließend mit der detaillierten Diskussion offener Fragen beginnen.

Die auffallendste und wichtigste Eigenschaft der Erythrocytenmembran ist deren selektive Anionenpermeabilität. Zu ihrer Erklärung hat Mond[38] — beeinflußt durch die Michaelisschen Untersuchungen über die Ionenpermeabilität der Collodiummembran — im Jahre 1927 folgende Hypothese aufgestellt. Die Erythrocytenmembran wird von wassergefüllten Poren durchzogen, deren Wandungen mit positiv geladenen, dissoziablen Festionen besetzt sind. Die Anionen können unter dem Gefälle ihres elektrochemischen Potentials durch diese Poren hindurchdiffundieren, während die Kationen aus elektrostatischen Gründen weitgehend zurückgehalten werden. Von verschiedenen Seiten[5, 7] wurde die Auffassung bestritten, daß die Festionen der Erythrocytenmembran dissoziabel sind. Es hat aber bisher niemand ernstlich bezweifelt, daß die selektive Anionenpermeabilität durch die Gegenwart positiver Festionen in wassergefüllten Kanälen erklärt werden muß, und ausführliche theoretische Betrachtungen, die von dieser Auffassung ausgehen, wurden wiederholt veröffentlicht, z. B. von WILBRANDT[67] u. SOLOMON[55, 56].

Die Selektivität der Erythrocytenmembran ist unvollständig. Durch Diffusion in Richtung des elektrochemischen Potentialgradienten wandern Kaliumionen aus dem Zellinneren ins Plasma, während Natriumionen in die Erythrocyten eindringen. Die Aufrechterhaltung einer vollständigen Impermeabilität für Kationen ist aber für das Überleben der Erythrocyten unerläßlich. Bei ungehindertem Fortschreiten der Kationenbewegungen würde es zu einer Wasseraufnahme durch die Erythrocyten kommen. Die Zellen müßten anschwellen und schließlich durch kolloidosmotische Hämolyse zugrunde gehen.

Um die Unvollkommenheit der elektrostatisch bedingten Selektivität auszugleichen, greift der Zellstoffwechsel in die Ionenpermeabilität ein. Er gleicht durch aktiven Transport die passive Wanderung der Alkalimetallionen aus und verhindert, daß durch chemische Analysen nachweisbare „Nettoverschiebungen" eintreten. Dank der Tätigkeit der Ionenpumpen verhält sich die Erythrocytenmembran so, als ob sie für K^+- und Na^+-Ionen undurchlässig wäre.

Die Bestandteile der Zellmembran, insbesondere ihre Phosphatide, unterliegen ständigem Stoffwechsel. Störungen des intermediären Zellstoffwechsels können daher nicht nur eine Hemmung des aktiven Transportes, sondern auch eine Änderung des Membranwiderstandes für die passiven Kationenbewegungen bewirken. Bei der Einregulierung der intracellulären Ionenkonzentration unterliegt somit nicht nur die Ionenakkumulation, sondern auch die passive Rückdiffusion der Kontrolle des Zellstoffwechsels.

Bei der Einstellung der stationären Nichtgleichgewichtsverteilung der Alkalimetallionen spielt der Stoffwechsel eine wesentliche Rolle, indem er den aktiven Transport bewerkstelligt. Obwohl gezeigt werden konnte, daß durch Stoffwechselgifte hervorgerufene abnorme Stoffwechselreaktionen zu Änderungen des Diffusionswiderstandes für die Alkalimetallionen führen können, ist es noch unklar, welche Bedeutung der Regulierung des Diffusionswiderstandes durch den Zellstoffwechsel unter physiologischen Bedingungen zukommt. Bei der Aufrechterhaltung der ungleichen Verteilung der Erdalkalimetallionen scheint dagegen die

Mitwirkung des aktiven Transportes von untergeordneter Bedeutung zu sein. Es ist aber sehr wahrscheinlich, daß der Zellstoffwechsel entscheidenden Anteil daran hat, die passive Penetration der Ca^{++}-Ionen aus dem Plasma in die Zellen zu verhindern.

Die nachfolgende Diskussion der Faktoren, welche die stationäre Ionenverteilung des Erythrocyten regeln, wird in zwei Unterabschnitte gegliedert:

1. Membranstruktur und selektive Anionenpermeabilität.
2. Stoffwechsel- und Kationenpermeabilität;
   a) aktiver Transport,
   b) Regulierung der passiven Rückdiffusion.

I. Selektive Anionenpermeabilität

Der Anionentransport durch die Erythrocytenmembran sei anhand neuerer Untersuchungen über die Sulfationenpermeabilität von Pferde- und Rindererythrocyten besprochen.

Sulfationenverschiebungen zwischen Suspensionsmedium und Zellinnerem finden — ebenso wie die der einwertigen Anionen — mit der Einstellung des Donnan-Gleichgewichtes ihr Ende. Es gilt:

$$\sqrt{\frac{SO_{4i}}{SO_{4a}}} = f_{Cl}\frac{Cl_i}{Cl_a} = f_{OH}\frac{OH_i}{OH_a} = r$$

f_{Cl} und f_{OH} sind Konstanten, die nur wenig von 1 verschieden sind. Die durch Gl. (1) beschriebene Relation wird bei beliebigen Werten

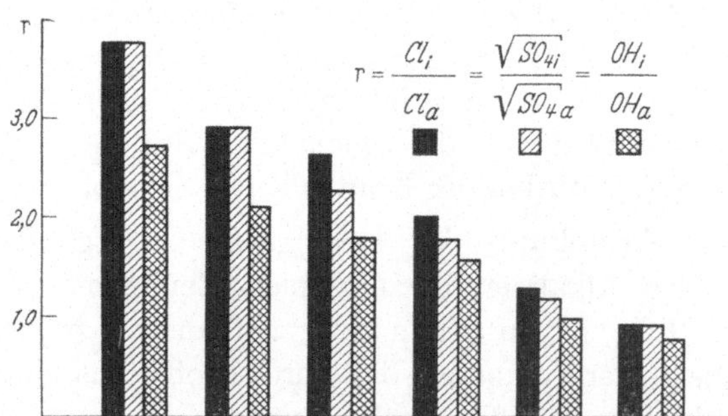

Abb. 1. Verteilung von Sulfat-, Chlor- und Hydroxylionen zwischen Zellinnerem und Außenlösung bei Variation der Gleichgewichtsbedingungen. Pferdeerythrocyten wurden in Lösungen suspendiert, die NaCl, Na_2SO_4 und Rohrzucker in wechselndem Mischungsverhältnis enthielten. Temperatur 32,5° C

des Donnan-Verhältnisses r beobachtet. Abbildung 1 stellt ein Versuchsbeispiel dar, bei dem r durch Variation des p_H und der

Salzkonzentration des Mediums verändert wurde. Die mit steigendem p_H zunehmende Dissoziation des Hämoglobins sowie der intracellulären Phosphorsäureester führt zu einer Verminderung von r und damit — wie die Abbildung zeigt — zu einer gleichsinnigen Verschiebung des Verteilungsverhältnisses aller im System vorhandenen Anionensorten. Die Einstellung des Donnan-Gleichgewichtes zeigt, daß es sich bei der Sulfationenpermeabilität ebenso wie bei der Permeabilität der sehr viel rascher penetrierenden einwertigen Anionen um Ionenbewegungen handelt, die „bergab", d. h. in Richtung des elektrochemischen Potentialgefälles verlaufen.

Es ist denkbar, daß auch beim passiven Transport vorübergehend chemische Reaktionen zwischen Membranbestandteilen und penetrierenden Anionen stattfinden. So hat man z. B. bei dem im Diffusionsgleichgewicht endenden Glucosetransport[30, 66] allen Grund zur Annahme, daß es sich um „bergab" gerichteten Carriertransport handelt. Die Sulfationen scheinen aber die Membran ohne Beteiligung eines Carriersystems zu durchqueren. So können zahlreiche Stoffwechselgifte wie Arsenat, Fluorid oder Monojodacetat die Sulfationenpermeabilität von Pferde- und Rindererythrocyten nicht beeinflussen. Ferner beobachteten wir, daß der Sulfationenfluß (mit $S^{35}O_4$ unter Gleichgewichtsbedingungen gemessen) proportional — und bei hohen Konzentrationen stärker als proportional — der Sulfationenkonzentration wächst[40]. Im Falle eines Carriertransportes hätte man statt dessen mit einer Sättigungskinetik rechnen müssen, d. h. man hätte erwarten müssen, daß bei hohen Sulfatkonzentrationen die Transportkapazität des Carriersystems voll ausgenutzt wird, so daß eine weitere Steigerung der Sulfationenkonzentration der Außenlösung keine Beschleunigung der Sulfationenpenetration mehr herbeiführen kann. Gegen das Vorhandensein eines Carriersystems spricht ferner, daß keine kompetitive Hemmung der SO_4-Permeabilität durch andere Anionensorten wie Chlorid, Lactat, Phosphat oder Chromat beobachtet werden konnte. Der recht geringe Einfluß aller dieser Ionensorten auf die Sulfationenpermeabilität kann in keinem Falle durch das Massenwirkungsgesetz beschrieben werden.

Von Villegas, Barton u. Solomon[63] wurde nachgewiesen, daß in der Erythrocytenmembran wassergefüllte Poren vorhanden sind.

GIEBEL u. PASSOW[16] konnten dann zeigen, daß hydrophile, organische Anionen auf ihrem Weg durch die Zellmembran durch diese wassergefüllten Poren diffundieren. Membranen mit wassergefüllten Poren können nur dann die Eigenschaft der selektiven Ionenpermeabilität besitzen, wenn die Porenwandungen elektrische Ladungen tragen. Man muß daher annehmen, daß in den Poren der Erythrocytenmembran positive Festionen vorhanden sind, welche die Kationen zurückhalten und den Anionen den Durchtritt gestatten.

Wir haben nun quantitative Untersuchungen über die Kinetik der Sulfationenpermeabilität von Rinder- und Pferdeerythrocyten ausgeführt*, um festzustellen, welche speziellen Annahmen über Konzentration, Anordnung und chemische Natur der Festionen gemacht werden müßten, um die selektive Anionenpermeabilität durch die Gegenwart fixierter, elektrischer Ladungen in wassergefüllten Membranporen erklären zu können. Bei der Planung der Experimente und der Diskussion der Resultate haben wir uns an der vor einigen Jahren von TEORELL[58] entwickelten und von SCHLÖGL[50] verallgemeinerten Theorie der Elektrolytdiffusion durch elektrisch geladene Membranen orientiert. In dieser Theorie wird die Ionenwanderung durch die konvektionsfreie wäßrige Phase einer symmetrischen Membran behandelt. Dabei wird angenommen, daß die Membran aus einem unregelmäßigen Maschenwerk besteht, in das vollständig dissoziierte Festionen eingebettet sind. Die Berechnung des Ionenflusses (ion flux) geht von folgenden Hypothesen aus:

1. Die Ionenkonzentrationen in den Membranoberflächen sind von den Ionenkonzentrationen der die Membran umgebenden Lösungen verschieden. Der Verteilungskoeffizient läßt sich unter der Annahme berechnen, daß sich — rasch im Vergleich zur Diffusion im Inneren der Membran — Donnan-Gleichgewichte an beiden Phasengrenzflächen einstellen.

2. Die Ionenbewegungen in der wäßrigen Phase des Membraninneren können im Prinzip ebenso berechnet werden wie bei Diffusionsvorgängen in einem großen Volumen wäßriger Lösung. Es ist daher erlaubt, für jede der gleichzeitig penetrierenden Ionensorten die Nernstsche Ionenbewegungsgleichung anzusetzen.

* Die Versuche wurden gemeinsam mit Drs. VIOLA PRIVAT u. H. LOHMANN ausgeführt.

3. Bei der Berechnung der sich im Inneren der Membran einstellenden Konzentrations- und Potentialgradienten darf angenommen werden, daß überall ein stationärer Zustand besteht.

Wir haben nun versucht zu überprüfen, innerhalb welcher Grenzen sich die der Theorie zugrunde liegenden Annahmen zur Interpretation der Kinetik der Anionenpermeabilität roter Blutkörperchen eignen. Beim Auftreten von Diskrepanzen zwischen Theorie und Experiment bemühten wir uns, die Theorie durch möglichst einfache Annahmen — etwa durch die Annahme, bis die Festionen dissoziabel sind — so weit zu modifizieren, bis eine befriedigende Übereinstimmung mit den Versuchsergebnissen erzielt wurde.

Mit unseren Versuchen wollten wir vor allem feststellen, ob die Sulfationenkonzentrationen in den Membranoberflächen — von denen die Penetrationsgeschwindigkeit in erster Linie abhängt — mit Hilfe des Donnan-Prinzips berechnet werden können (Annahme 1 der Teorell-Theorie). Falls es sich nämlich bei der Sulfationenverteilung an den Phasengrenzflächen um Donnan-Gleichgewichte handeln sollte, wäre es möglich, die Konzentration der die Gleichgewichtslage bestimmenden Festionen zu berechnen und ihre Dissoziationskonstante zu ermitteln.

Wir haben zunächst eine auf unsere Versuchsbedingungen anwendbare Gleichung abgeleitet, um die Sulfationenkonzentration in der Zellmembran berechnen zu können. Dabei durften wir voraussetzen, daß unter unseren Versuchsbedingungen nur Cl^-, OH^- und SO_4^{--} als diffusible Anionen vorhanden sind. Wir nehmen nun an, daß an den beiden Membrangrenzflächen Donnan-Gleichgewichte für alle diffusiblen Anionensorten existieren. Im Innern der Zellmembran gilt die Elektroneutralitätsbedingung: Die Summe der negativen Ladungen der diffusiblen Anionen ist gleich der Zahl fixierter Kationen. Für den Fall, daß die Festionen dissoziabel sein sollten (z. B. NH_3^+-Gruppen könnten in $NH_2 + H^+$ dissoziieren), muß ferner das Massenwirkungsgesetz berücksichtigt werden. Durch Kombination dieser drei Bedingungen, nämlich Donnan-Gleichgewichte an den Phasengrenzflächen (Gl. 1), Elektroneutralität im Innern der Zellmembran (Gl. 2) und Anwendbarkeit des Massenwirkungsgesetzes zur Berechnung des Dissoziationsgrades der Festionen (Gl. 3), erhält man die unten dargestellte Formel zur Berechnung der Sulfationenkonzentration in der Zellmembran.

$$-A^+ + 2\,SO_{4m} + Cl_m = 0 \qquad \text{(Elektroneutralitätsbedingung)} \qquad (1)$$

$$A^+ = \bar{A} \cdot \frac{H_m}{K + H_m} \qquad \text{(Dissoziationsgleichgewicht der Festionen)} \qquad (2)$$

$$\sqrt{\frac{(SO_{4a})}{(SO_{4m})}} = \frac{(Cl_a)}{(Cl_m)} = \frac{(H_m)}{(H_a)} \qquad \text{(Donnan-Gleichgewicht zwischen Membran und angrenzenden Lösungen)} \qquad (3)$$

Erklärung der Symbole:

A^+ = Festionenkonzentration,

$\bar{A}$ = Konzentration dissoziierter und undissoziierter Aminogruppen,

K = Massenwirkungskonstante.

Die Indices m und a bezeichnen Konzentrationen in der Zellmembran und in der Außenlösung. Die Klammern bezeichnen Aktivitäten, während chemische Symbole ohne Klammern die Bedeutung von Konzentrationen haben.

Zur Abkürzung werden folgende Definitionen verwendet:

$$p = \frac{(Cl_a)}{\sqrt{(SO_{4a})}} = \frac{(Cl_m)}{\sqrt{(SO_{4m})}}$$

$$q = (H_a) \cdot \sqrt{(SO_{4a})} = (H_m) \cdot \sqrt{(SO_{4m})}\,.$$

Durch Kombination der Gl. (1), (2) und (3) erhält man unter Verwendung der beiden Definitionen folgende Gleichung:

$$2\,SO_{4m} + f \cdot p \cdot \sqrt{SO_{4m}} - \bar{A} \cdot \frac{q}{K \cdot \sqrt{SO_{4m}} + q} = 0 \qquad (4)$$

mit $f = \frac{\sqrt{f\,SO_{4m}}}{f\,Cl_m}$, wenn $f\,SO_{4m}$ und $f\,Cl_m$ die individuellen Ionenaktivitätskoeffizienten für Sulfat- und Chlorionen in der Zellmembran bedeuten.

Einige Konsequenzen aus den geschilderten Annahmen (bzw. Gl. 4) lassen sich qualitativ verständlich machen.

1. Wenn bei konstanter Sulfationenkonzentration die Chlorionenkonzentration vergrößert wird, so muß die Sulfationenkonzentration in der Zellmembran geringer werden. Denn nun konkurrieren die Chlorionen mit den Sulfationen um die kationischen Äquivalente der Festionen.

2. Wenn man annimmt, daß es sich bei den Festionen um dissoziable Aminogruppen handelt, so ist zu erwarten, daß mit zunehmendem p_H die Konzentration positiver Festionen abnimmt. Die Verminderung positiver Ladungen in der Zellmembran muß zu einem Absinken der Konzentration diffusibler Anionen führen.

Wir nehmen nun an, daß der Sulfationenfluß (flux, gemessen in Mol/cm²/sec) proportional der Konzentration der Sulfationen in

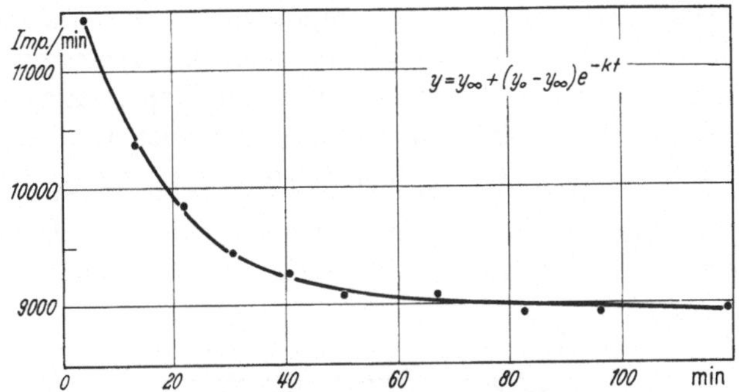

Abb. 2. Zeitlicher Verlauf des Eindringens von $S^{35}O_4$ in Rindererythrocyten. Die Zellen wurden vor Zusatz der Radioaktivität mit einer isotonischen Mischung von Na_2SO_4-NaCl äquilibriert. SO_4-Konzentration: 32 mMol/l. p_H: 7,4 (Gleichgewichtskonzentrationen in der Außenlösung). Ausgezogene Kurve berechnet mit Hilfe der Methode der kleinsten Quadrate. $y_0 = 12229$, $y_{00} = 8967$, $K = 0{,}061$
Aus der Konstanten K werden Ionenfluß und Permeabilitätskonstante berechnet. Ordinate: Impulszahl in der Außenlösung. Abszisse: Zeit in Minuten

der Zellmembran ist. Die Sulfationenkonzentration in der Zellmembran läßt sich nun annähernd* mit Hilfe von Gl. 4 berechnen. Man kann daher prüfen, ob den Änderungen der Sulfationenkonzentration in der Zellmembran, die bei experimenteller Variation von p_H und Cl^--Konzentration der Außenlösung zu erwarten sind, Änderungen des Ionenflusses parallel gehen.

Diese einfache Überlegung gilt aber nur dann, wenn es gelingt, eine Schwierigkeit auszuschalten. Bei allen Ionenverschiebungen treten Diffusionspotentiale auf, die auf die Penetrationsgeschwindigkeit rückwirken und deren Größe an der Erythrocytenmembran bisher noch nicht gemessen werden konnte. Wir waren daher darauf angewiesen, unsere Versuchsanordnung so zu wählen, daß praktisch keine Diffusionspotentiale auftreten können, so daß bei

* Die exakte Auswertung von Gl. 4 ist vor allem deshalb unmöglich, weil die Aktivitätskoeffizienten in der Zellmembran unbekannt sind.

einer Änderung des Ionenmilieus lediglich der Verteilungskoeffizient variiert wird. Diese Forderung läßt sich in einfacher Weise verwirklichen, indem die gewaschenen Erythrocyten mit isotonischen, natriumsulfathaltigen Lösungen so lange äquilibriert werden, bis sich das Donnan-Gleichgewicht eingestellt hat, so daß die Diffusionspotentiale verschwinden. Anschließend wird eine unwägbare Menge an radioaktivem Sulfat zur Außenlösung hinzugefügt und das Eindringen der Radioaktivität verfolgt. Der zeitliche Verlauf des Absinkens der Impulszahl in der Außenlösung läßt sich, wie Abb. 2 zeigt, durch eine einfache Exponentialfunktion beschreiben. Aus der Konstanten im Exponenten können Ionenfluß und Permeabilitätskonstante berechnet werden.

Abb. 3 zeigt nun die p_H-Abhängigkeit von Permeabilitätskonstanten, die unter diesen Versuchsbedingungen gemessen wurden. Sie gibt eine Bestätigung älterer Mondscher Befunde[38], indem

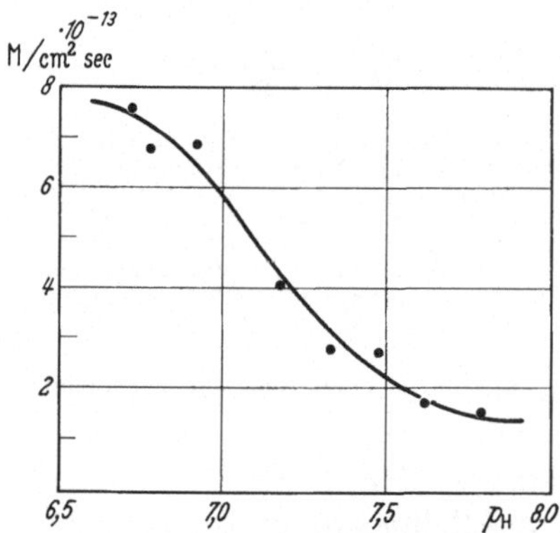

Abb. 3. p_H-Abhängigkeit der Sulfationenpermeabilität im Donnan-Gleichgewicht. Die Rindererythrocyten wurden vor Zusatz der Radioaktivität mit Na_2SO_4-NaCl-Mischungen von verschiedenem p_H in das Donnan-Gleichgewicht gebracht. p_H-Einstellung mit Phosphatpuffer. SO_4-Konzentration: 30 mMol/l. Cl-Konzentration: 32 mMol/l Gleichgewichtskonzentrationen in der Außenlösung). Temperatur: 32,5° C. Ordinate: Ionenfluß (berechnet nach dem in Abb. 2 dargestellten Verfahren). Abszisse: p_H der Außenlösung (Gleichgewichtswert)

sie zeigt, daß die Penetrationsgeschwindigkeit beim Übergang von p_H 6,8 auf 7,8 auf rund $^1/_3$ bis $^1/_4$ absinkt. Der Wendepunkt dieser Kurve liegt zwischen p_H 7,1 und 7,2. Dieser Wendepunkt ist nicht notwendigerweise ein unmittelbares Maß für die Dissoziationskonstante von Festionen, da der Verteilungskoeffizient für Hydroxyl- bzw. Wasserstoffionen zunächst unbekannt ist.

Die Steigerung der Penetrationsgeschwindigkeit mit abnehmendem p_H läßt sich durch die Annahme erklären, daß die Festionenkonzentration im sauren Bereich größer wird als im alkalischen*.

Durch Zusatz geeigneter Mengen an Chlorionen sollte es möglich sein, die Grenzflächenkonzentration der Sulfationen trotz Senkung des p_H konstant zu halten oder sogar zu vermindern. Die Chlorionen würden unter diesen Bedingungen die bei p_H-Senkung neugebildeten Festionen elektrisch absättigen. Falls diese

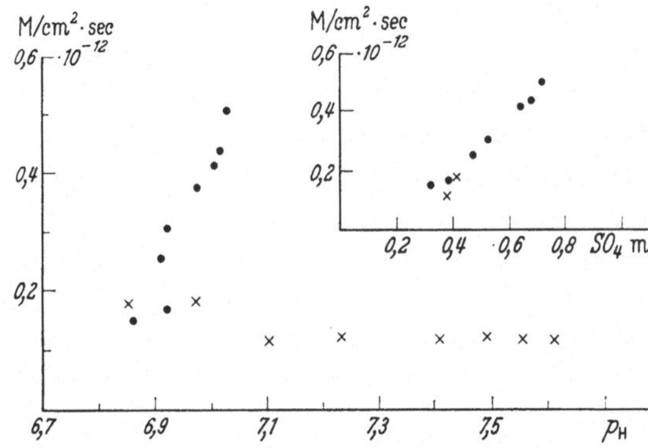

Abb. 4. Abhängigkeit der Permeabilitätskonstanten von der Lage des Donnan-Gleichgewichtes. Pferdeerythrocyten wurden in isotonischen Lösungen suspendiert, die NaCl, Na_2SO_4 und Rohrzucker in wechselndem Mischungsverhältnis enthielten. Temperatur: 32,5° C. Ordinate: Ionenfluß (berechnet nach dem in Abb. 2 dargestellten Verfahren). Abszisse: p_H der Außenlösung (Gleichgewichtswert). Einsatz: Experimentelle Prüfung von Gl. (4). Die beiden Kreuze repräsentieren die beiden Extremwerte des der Abszisse in Abb. 4 parallel laufenden Kurvenastes (ausgefüllte Punkte). Die dazwischenliegenden Werte füllen den Raum zwischen den Kreuzen so eng aus, daß auf eine Darstellung verzichtet wurde. Zur Berechnung der Grenzflächenkonzentration SO_{4m} (Abszisse) wurde für die maximale Festionenkonzentration $\bar{A}$ ein Wert von 3 Mol/l und für die Dissoziationskonstante der Festionen K ein Wert von $1 \cdot 10^{-9}$ (Aminogruppen) eingesetzt. Ordinate: Ionenfluß (berechnet nach dem in Abb. 2 dargestellten Verfahren)

Überlegung richtig ist, sollte es also möglich sein, durch Zusatz geeigneter Mengen an Chlorionen die p_H-Abhängigkeit zu unterdrücken oder gar umzukehren. Abb. 4 zeigt, daß sich diese

* Wir gehen hier von der Annahme aus, daß sich nur der Verteilungskoeffizient und nicht der Diffusionswiderstand mit dem p_H ändert. Es ist unwahrscheinlich, daß diese Annahme ohne Einschränkung gilt. So ist es denkbar, daß eine mit einer p_H-Steigerung verbundene Verminderung der Membranladung zu einer Abnahme des Porenradius führt (Ussing, persönl. Mitteilung). Ferner müßte berücksichtigt werden, daß die Größe des Selbstdiffusionskoeffizienten von der Ionenstärke in der Membran abhängt, die ihrerseits eine Funktion der Festionenkonzentration sowie des Mischungsverhältnisses Cl/SO_4 ist.

Voraussage experimentell bestätigen läßt. Je nach Wahl des Verhältnisses H^+/Cl^- lassen sich beliebige Werte für den Sulfationenfluß erzielen.

Mit Hilfe von Gl. 4, in der die hier qualitativ formulierten Gedankengänge enthalten sind, ist es nun möglich, sämtliche Meßergebnisse auf einen einheitlichen Nenner zu bringen. Unter Annahme eines vernünftigen Wertes für die Festionenkonzentration (3 Mol/l) und unter Benutzung des Wertes für die Dissoziationskonstante für Aminogruppen (1×10^{-9}) gelingt es zu zeigen, daß der Sulfationenfluß — dessen p_H-Abhängigkeit nach Belieben verschoben werden kann (Abb. 4) — der berechneten Sulfationenkonzentration in der Zellmembran parallel geht. Der Einsatz in Abb. 4 zeigt nun, daß die Punkte, die in Abb. 4 auf zwei verschiedenen Kurven lagen, auf eine gemeinsame Gerade fallen. Mit allem Vorbehalt, den man der unter einer begrenzten Zahl von Versuchsbedingungen gefundenen Übereinstimmung zwischen Meßergebnissen und Theorie entgegenbringen muß, läßt sich somit schließen, daß Aminogruppen in einer Konzentration von etwa 3 Mol/l die selektive Anionenpermeabilität der Erythrocytenmembran bedingen.

Mit denselben Vorbehalten, mit denen man Angaben über Art und Konzentration der Festionen betrachten muß, kann man annehmen, daß zwischen Membranoberflächen und angrenzenden Lösungen ein Donnan-Gleichgewicht für die Anionen vorliegt und daß die Einstellung dieses Gleichgewichtes sehr viel rascher erfolgt als der Diffusionsvorgang im Innern der Membran.

Die zweite Voraussetzung der Teorellschen Theorie — nämlich die Anwendbarkeit der Nernstschen Ionenbewegungsgleichung für alle penetrierenden Ionensorten — kann dagegen wahrscheinlich nicht erfüllt sein. Diese Feststellung stützt sich auf folgende Beobachtungen und Überlegungen.

Wenn es erlaubt ist, die Anionenverteilung an den Membrangrenzflächen als Donnan-Gleichgewicht aufzufassen, so läßt sich der im Membraninneren geltende Selbstdiffusionskoeffizient D für die Sulfationen berechnen* Nimmt man an, daß die effektive

* In einer homogenen Membran ist

$$D = \frac{\text{Ionenfluß — Membrandicke — 100}}{SO_{4m} \cdot \text{effektive Diffusionsfläche (in \%)}}$$

Diffusionsfläche nur rund 0,5‰ der gesamten Membranoberfläche ist und die Membrandicke 100 Å beträgt, so erhält man D 1×10^{-13} cm²/sec. Vergleicht man diesen Wert mit Literaturangaben für die Selbstdiffusionskoeffizienten zweiwertiger Ionen in stark vernetzten Ionenaustauscherharzen[54], so ergibt sich, daß der von uns gefundene Selbstdiffusionskoeffizient rund 10^5mal kleiner ist als der niedrigste Wert, der bisher in synthetischen Materialien gemessen wurde.

Der Dissoziationsgrad von Aminogruppen nimmt bei Temperatursteigerung zu. Sollte es sich bei den Festionen tatsächlich um Aminogruppen handeln, so müßte man annehmen, daß ihre Konzentration und damit auch die der diffusiblen Sulfationen im Innern der Membran mit zunehmender Temperatur kleiner wird.

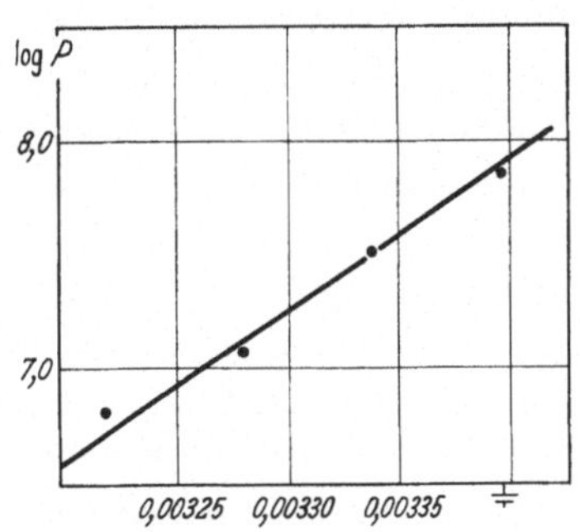

Abb. 5. Temperaturabhängigkeit der Sulfationenpermeabilität. Pferdeerythrocyten. pH 7,4. Ordinate: Logarithmus der Permeabilitätskonstanten. Abscisse: Kehrwert der absoluten Temperatur

Eine durch Temperatursteigerung bedingte Zunahme des Selbstdiffusionskoeffizienten sollte daher nicht mit einer entsprechenden Zunahme des mit $S^{35}O_4$ gemessenen Ionenflusses einhergehen; denn die Vergrößerung der Penetrationsgeschwindigkeit der einzelnen Ionen müßte mehr oder weniger weitgehend durch die Verminderung der Zahl penetrierender Ionen kompensiert werden, so daß man nur einen niedrigen Temperaturkoeffizienten für den Sulfationenfluß erwarten würde. Wir beobachten aber eine sehr große scheinbare Aktivierungsenergie von 28 bis 30 Kcal pro Grad[8,40] (Abb. 5). Aus dem Absolutwert des Diffusionskoeffizienten und dem Wert für die Aktivierungswärme des Sulfationenflusses (der aus der Aktivierungsenergie berechnet werden kann) errechnet sich eine außergewöhnlich hohe positive Aktivierungsentropie, etwa 90 cal/Grad. Sie besitzt dieselbe Größenordnung wie die bei der Eiweißdenaturierung beobachteten Aktivierungsentropien.

Da bei Versuchen mit Ionenaustauschharzen niemals höhere Aktivierungsenergien als etwa 16 Kcal/Grad gefunden wurden[54], könnte die große beobachtete Aktivierungsenergie als wesentliches Argument gegen die Brauchbarkeit der Teorellschen Fixed Charge Theory angesehen werden. Man sollte aber die Möglichkeit in Betracht ziehen, daß sich die Membranstruktur mit zunehmender Temperatur „auflockert“, so daß der Diffusionswiderstand abnimmt. Man darf ferner daran denken, daß die Erythrocytenmembran eine kompaktere Struktur besitzen könnte als Ionenaustauscherharze, so daß die elektrostatischen Anziehungskräfte zwischen Festionen und penetrierenden Anionen sowie die Abstoßungskräfte zwischen sich nähernden diffusiblen Anionen abnorm hohe Beträge zur Aktivierungsenergie liefern können.

Bei der Eiweißdenaturierung werden Wasserstoffbrückenbindungen aufgesprengt. Man nimmt an, daß diese Reaktionen für die große Entropieänderung beim Aktivierungsprozeß verantwortlich gemacht werden müssen. Es ist denkbar, daß bei der Diffusion der Anionen durch die Erythrocytenmembran eine Deformierung der Poren stattfindet. Dabei könnte es vielleicht zur vorübergehenden Spaltung von Wasserstoffbrücken kommen. Ein derartiger Vorgang würde zwanglos die hohe positive Aktivierungsentropie des Penetrationsprozesses erklären können.

Wenn man nicht annehmen möchte, daß der so niedrige Wert für den Selbstdiffusionskoeffizienten und der abnorme Wert für die Aktivierungsentropie durch fehlerhafte Annahmen bei der Berechnung der Verteilungskoeffizienten zwischen Suspensionsmedium und Membrangrenzflächen zustande kommen (d. h., wenn man nicht annehmen möchte, daß es unzulässig ist, die Sulfationenkonzentration in der Zellmembran mit Hilfe der Donnan-Theorie zu berechnen), so deuten diese Beobachtungen und Überlegungen darauf hin, daß entweder der Verteilungskoeffizient von falschen Voraussetzungen ausgehend berechnet wurde oder daß der Diffusionsmechanismus der Sulfationen in den engen Poren der Zellmembran sich grundsätzlich vom Diffusionsmechanismus in einem großen Volumen wäßriger Lösung unterscheidet. Hierfür sprechen folgende Argumente:

Die Anionen penetrieren rund 5×10^5mal rascher durch die Erythrocytenmembran als die Kationen. SOLOMON[56] hat die Festionenkonzentration berechnet, die erforderlich wäre, um die großen Unterschiede der Penetrationsgeschwindigkeiten von An- und Kationen zu erklären, wenn die Ionenbeweglichkeiten in der

Zellmembran genauso groß wären wie in wäßriger Lösung. Seine Rechnung ergab einen Wert von über 100 Mol/l Membransubstanz. Die Annahme einer derartig hohen Festionenkonzentration wäre absurd. Man darf daher vermuten, daß die Ionenbeweglichkeiten im Innern der Zellmembran sehr stark vermindert sind, und man kann daran denken, daß eine räumliche Behinderung des Diffusionsvorganges in den Membranporen erfolgt. Vergleicht man den von VILLEGAS u. Mitarb.[63] sowie von GIEBEL u. PASSOW[16] übereinstimmend mit etwa 4 Å angegebenen Porenradius mit dem Radius der penetrierenden Sulfationen, so ergibt sich, daß es für ein aus der Außenlösung kommendes Sulfation unmöglich sein dürfte, mit einem in einer Pore befindlichen Sulfation den Platz zu wechseln. Die Poren sind wahrscheinlich so eng, daß die beiden gleichsinnig geladenen und sich daher abstoßenden Ionen nicht aneinander vorbei können* und die Nernstsche Ionenbewegungsgleichung — zumindest zur Beschreibung von Anionenverschiebungen, die vom Nichtgleichgewichtszustand ins Donnan-Gleichgewicht führen — nicht mehr ohne weiteres angewandt werden darf.

Unter diesen Umständen kann eine Penetration nur nach dem single file Mechanismus (KEYNES u. HODGKIN[24]) oder einem ähnlichen Prinzip erfolgen. Es ist durchaus denkbar, daß bei der Penetration durch sehr enge Poren, bei der die in entgegengesetzten Richtungen diffundierenden Ionen sich gegenseitig behindern, extrem niedrige Selbstdiffusionskoeffizienten und sehr hohe scheinbare Aktivierungsenergien auftreten.

In der Teorell-Theorie wird eine gegenseitige Behinderung entgegengerichteter Ionenströme nicht berücksichtigt. Aber auch beim single file Mechanismus muß die mit Isotopen unter *Gleichgewichtsbedingungen* gemessene Penetrationsgeschwindigkeit von der SO_4-Konzentration in der Membran abhängen. Die Berechnung dieser Konzentration mit Hilfe der Donnan-Gesetze ist daher selbst dann sinnvoll, wenn die anderen Voraussetzungen der Teorell-Theorie nicht zutreffen sollten. Es muß allerdings offen bleiben, ob bei einem single file Mechanismus Proportionalität zwischen Ionenfluß und Sulfationenkonzentration in den Poren besteht.

Zusammenfassend läßt sich feststellen, daß es möglich ist, die Untersuchungen über die Kinetik der Sulfationenpermeabilität

* Es sei denn, daß die Poren deformierbar sind.

durch die Annahme zu erklären, daß in den die Erythrocytenmembran durchziehenden wassergefüllten Poren positive Festionen vorhanden sind. Die p_H-Abhängigkeit der Sulfationenpermeabilität deutet darauf hin, daß es sich bei den Festionen um Aminogruppen in einer Konzentration von 3 Mol/l handeln könnte. Mit denselben Vorbehalten, die man den Angaben über Natur und Konzentration der Festionen entgegenbringen muß, darf man annehmen, daß sich — rasch im Vergleich zur Penetration — an den Membrangrenzflächen Donnan-Gleichgewichte einstellen. Auf die Ionenbewegungen im Inneren der Membran lassen sich die einfachen Gesetze der Elektrolytdiffusion (Nernstsche Ionenbewegungsgleichung) wohl nicht anwenden, da die Ionen sich auf ihrer Wanderung durch die Membran in den engen Poren gegenseitig behindern dürften.

Der vorangehenden Diskussion wurden ausschließlich die Modellvorstellungen der Teorellschen Fixed Charge Theory zugrunde gelegt. Überall, wo Schwierigkeiten auftraten, wurde zu zeigen versucht, in welcher Hinsicht die Theorie abgewandelt werden muß, um den Beobachtungsergebnissen gerecht zu werden. Dabei wurde offen gelassen, ob es möglich ist, unsere Daten auch durch Theorien zu deuten, die auf gänzlich andersartigen Voraussetzungen basieren. Diese bewußte Einseitigkeit schien mir aber gerechtfertigt, da die Fixed Charge Theory von den einfachst möglichen Annahmen über die Natur der selektiven Ionenpermeabilität ausgeht und es wünschenswert sein dürfte festzustellen, wie weit sie zur Deutung von Eigenschaften der Erythrocytenmembran ausreichen.

II. Zellstoffwechsel und Kationenpermeabilität

a) Aktiver Transport der Alkalimetallionen. Die Erythrocytenmembran ist für Kalium- und Natriumionen durchlässig[48, 53], dennoch werden ständig große Konzentrationsunterschiede zwischen Zellinnerem und Außenlösung aufrechterhalten. So beträgt beispielsweise bei Menschenerythrocyten die Kaliumkonzentration des Zellwassers etwa 150 mMol/l, die des Serums nur 4,5 mMol/l. Die Natriumkonzentration ist im Serum mit etwa 145 mMol/l wesentlich höher als im Erythrocyteninneren, wo sie nur rund 10 mMol/l Zellwasser beträgt. Diese ungleiche Verteilung der Alkalimetallionen beruht nicht auf der Einstellung eines Donnan-

Gleichgewichtes. Falls nämlich ein Donnan-Gleichgewicht existieren würde, müßte folgende Beziehung gelten*:

$$\frac{K_i^+}{K_a^+} = \frac{Na_i^+}{Na_a^+} = \frac{Cl_a^-}{Cl_i^-}\,.$$

Das aus der Chlorionenverteilung zwischen Plasma- und Zellwasser errechenbare Donnan-Verhältnis Cl_a/Cl_i nimmt bei p_H 7,4 etwa den Wert 1,4:1 an. Es unterscheidet sich damit größenordnungsmäßig von den Verteilungsquotienten für Kalium- und Natriumionen, die bei 1:30 bzw. 1:15 liegen. Es handelt sich also bei der ungleichen Verteilung der Alkalimetallionen um einen Nichtgleichgewichtszustand, bei dem „bergab" verlaufende Ionenverschiebungen durch „bergauf" gerichteten aktiven Transport kompensiert werden.

Wird der Stoffwechsel der Erythrocyten durch Temperatursenkung vermindert, so erfolgen Kaliumverlust und Natriumaufnahme. Diese Ionenverschiebungen können nach Wiedererwärmen und Zusatz von Substraten rückgängig gemacht werden, wobei die Kalium- und Natriumionen dann entgegen dem Gefälle des elektrochemischen Potentials wandern[22]. Als energieliefernde Substrate für diese Akkumulationsvorgänge kommen in erster Linie Glucose und Nucleoside, wie Adenosin oder Inosin, in Frage[37]. Der Umsatz dieser Substrate erfolgt bei den reifen, kernlosen Erythrocyten fast ausschließlich durch anaerobe Stoffwechselvorgänge, bei denen Milchsäure als Stoffwechselprodukt entsteht. Die Glucose wird dabei durch die üblichen glucolytischen Reaktionen umgesetzt, während die Nucleoside nach phosphorylytischer Spaltung Pentosephosphat liefern[11]. Das Pentosephosphat wird über den Pentosephosphatcyclus in Triosephosphat[2a] überführt, welches dann am weiteren Verlauf der glucolytischen Reaktionen teilnimmt. Glucolysegifte, wie NaF oder Monojodessigsäure, hemmen den aktiven Transport, während Atmungsgifte, wie Cyanid, Acid oder Dinitrophenol, unwirksam sind[10].

In den folgenden Abschnitten sollen nun einige neuere Untersuchungen über die Beziehungen zwischen Stoffwechsel und aktivem Transport geschildert werden. Dabei werden aus der Fülle

* Hier kennzeichnen die Indices *i* und *a* die Ionenkonzentrationen im Zellinneren und in der Außenlösung.

des vorhandenen Materials nur solche Befunde herausgegriffen, die eine ausführliche Diskussion der folgenden Fragen ermöglichen:

1. Wieviel Energie muß durch den Zellstoffwechsel zur Verfügung gestellt werden, um die stationäre Kationenverteilung dauernd aufrecht zu erhalten?

2. Durch welche Stoffwechselreaktionen wird der aktive Transport bewerkstelligt?

Der Energiebedarf zur Aufrechterhaltung der stationären Kationenverteilung

Die Größenordnung des von den Erythrocyten zur Aufrechterhaltung der ungleichen Kationenverteilung aufzuwendenden Energiebetrages läßt sich abschätzen. Dabei wird zunächst der Arbeitsbetrag ermittelt, der im stationären Zustand zur Überführung eines gr-Äquivalentes einer Ionensorte aus dem Plasma ins Zellwasser aufgebracht werden muß. Er beträgt mindestens

$$A = RT \ln \frac{a_{\mathrm{I}}}{a_{\mathrm{II}}} + zF\Delta\varphi \,. \tag{1}$$

A = Arbeitsbetrag
R = Gaskonstante
T = absolute Temperatur
$a_{\mathrm{I}}, a_{\mathrm{II}}$ = Ionenaktivitäten in Plasma- und Zellwasser
z = Wertigkeit
F = Faraday
$\Delta\varphi$ = Membranpotential

Der erste Summand dieser Gleichung gibt die osmotische, der zweite die elektrische Überführungsarbeit an. Bisher konnte noch keine der Größen a_{I}, a_{II}, $\Delta\varphi$ direkt bestimmt werden. Anstelle des Verhältnisses von Ionenaktivitäten $a_{\mathrm{I}}/a_{\mathrm{II}}$ zwischen Außenlösung und Erythrocyteninnerem muß daher das Verhältnis der Ionenkonzentrationen $c_{\mathrm{I}}/c_{\mathrm{II}}$ treten. Direkte Messungen des Membranpotentials $\Delta\varphi$ scheinen bisher nicht gelungen zu sein. Die Größe des Membranpotentials muß daher aus dem Donnan-Verhältnis für die Chlorionenverteilung berechnet werden, wobei wiederum mangels genauerer Kenntnisse anstelle der Ionenaktivitäten die Ionenkonzentrationen $\frac{Cl_{\mathrm{I}}}{Cl_{\mathrm{II}}}$ verwendet werden müssen. Man erhält auf diese Weise für $\Delta \varphi = R \cdot T \cdot \ln \cdot \frac{Cl_{\mathrm{I}}}{Cl_{\mathrm{II}}}$ einen Wert von 6 bis 7 mV (p_H 7,4).

Zur Berechnung der von der Zelle beim aktiven Transport vollbrachten Arbeitsleistung muß A mit der pro Zeiteinheit entgegen dem Gefälle des elektrochemischen Potentials bewegten Ionenmenge M_a multipliziert werden. Die Größe M_a wird im allgemeinen dem mit radioaktiven Isotopen im stationären Zustand gemessenen Ionenfluß gleichgesetzt. M_a läßt sich aber nur dann zuverlässig abschätzen, wenn der Transportmechanismus soweit bekannt ist, daß vom gesamten mit Hilfe des Isotops gemessenen Ionenfluß der Anteil abgezogen werden kann, der thermodynamisch „erlaubt" ist. Es muß beispielsweise damit gerechnet werden (USSING 1949[62], 1960[62a]), daß ein Teil der Ionenbewegungen durch die Membran hindurch durch sogenannte Austauschdiffusionen (exchange diffusion) erfolgen kann. Dabei kann ein großer Teil der Energie für die Ionenbewegungen entgegen dem Gefälle des elektrochemischen Potentials durch den gleichzeitigen Ionenfluß derselben Ionensorte in Richtung des Gefälles gedeckt werden. So wurde von TOSTESON u. ROBERTSON[60] bei Vogelerythrocyten rasche Austauschdiffusion beobachtet. Sie spielt aber anscheinend bei den Erythrocyten der Säugetiere nur eine sehr geringe Rolle.

Folgende Überlegung zeigt, daß selbst dann, wenn Austauschdiffusion ausgeschlossen werden kann, der mit Isotopen gemessene Ionenfluß immer noch nicht ohne weiteres zur Berechnung der beim aktiven Transport zu vollbringenden Arbeitsleistung benutzt werden darf.

Im stationären Zustand ist der aktive Ionentransport M_a ebenso groß wie die von II nach I gerichtete passive Ionenbewegung M_p:

$$M_a - M_p = 0 . \tag{2}$$

M_p kann als Differenz zweier entgegengerichteter Ionenflüsse m_{I} und m_{II} aufgefaßt werden.

$$M_p = m_{\mathrm{II}} - m_{\mathrm{I}} . \tag{3}$$

Wenn die in entgegengesetzten Richtungen wandernden Ionen sich in der Membran nicht gegenseitig beeinflussen, ist m_{II} proportional c_{II} und m_{I} proportional $c_{\mathrm{I}} \cdot \zeta$

$$\left(\zeta = e^{-\frac{zF \cdot \Delta \varphi}{RT}}\right) .$$

Die Größe des Proportionalitätsfaktors K hängt dabei von den Versuchsbedingungen ab.

In Fällen, in denen die Goldmannsche Gleichung auf die passive Rückdiffusion der Elektrolyte angewendet werden darf, gilt beispielsweise (vgl. [24a])

$$K = \frac{K^*}{1-\xi}.$$

K ist hier eine Konstante, welche den Membraneigenschaften und den Eigenschaften der diffundierenden Ionen Rechnung trägt.

Fügt man radioaktives Kalium zu einer Erythrocytensuspension hinzu, so wird ein Teil der Radioaktivität, nämlich

$$M_p = M_a = K \cdot (c_{\mathrm{II}} - c_{\mathrm{I}} \cdot \zeta) \tag{4}$$

mit Hilfe der Ionenpumpe und der Rest durch passive Ionenbewegung aus der Außenlösung (I) ins Zellinnere (II) eindringen. Der gesamte mit radioaktiven Isotopen gemessene Ionenfluß ist also um den Betrag $K \cdot c_{\mathrm{I}} \cdot \zeta$ größer als die durch die Ionenpumpe geförderte Kaliummenge. Für die von der Zelle pro Zeiteinheit geleistete Arbeit Å ergibt sich somit:

$$Å = m_{\mathrm{II}} \cdot \left(1 - \frac{c_{\mathrm{I}}}{c_{\mathrm{II}}} \cdot \zeta\right) \left(R\,T \ln \frac{c_{\mathrm{I}}}{c_{\mathrm{II}}} + z\,F \Delta\varphi\right). \tag{5}$$

Wenn der Abstand des Systems vom Donnan-Gleichgewicht sehr groß ist, kann der thermodynamisch „erlaubte" (d. h. passive) Anteil des mit Isotopen gemessenen Ionenflusses ($K \cdot c_{\mathrm{I}} \cdot \zeta$) neben dem durch aktiven Transport bewerkstelligten Ionenfluß vernachlässigt werden. Unter diesen Umständen degeneriert die obige Gleichung zu einem Ausdruck, der von USSING[62], SOLOMON[55] u. a. schon vor langer Zeit verwendet wurde:

$$Å = m_{\mathrm{II}} \cdot \left(R\,T \ln \frac{c_{\mathrm{I}}}{c_{\mathrm{II}}} + zF \Delta\,\varphi\right). \tag{6}$$

Die Verteilung der Alkalimetallionen zwischen Serum und Menschenerythrocyten ist so weit vom Donnan-Gleichgewicht entfernt, daß die Anwendung dieser einfacheren Gleichung zur Durchführung einer Überschlagsrechnung gerechtfertigt ist. SOLOMON[55] berechnete mit ihrer Hilfe, daß Menschenerythrocyten etwa 9 cal/l Zellen/Std. aufwenden müssen, um bei einem Na^+-Ionenfluß von 3,1 mäquiv./l Zellen/Std. und einem K^+-Ionenfluß von 1,7 mäquiv./l Zellen/Std. die ungleichmäßige Kationenverteilung aufrechtzuerhalten. Menschenerythrocyten setzen etwa 2,3 mMol Glucose/l Zellen/Std. um. Dabei werden durch die Glucolyse rund 110 cal/l Zellen/Std. freigesetzt. Es werden also nur rund 8% der Stoffwechselenergie für den aktiven Transport benötigt.

Der Mechanismus des aktiven Ionentransportes

Über den Mechanismus der beim aktiven Ionentransport erfolgenden Umwandlung von Stoffwechselenergie in osmotische und elektrische Arbeit konnte man bis vor etwa einem Jahr nur sehr allgemeine Aussagen machen. Man mußte sich weitgehend auf die Diskussion der physikochemischen Voraussetzungen, unter denen Transport entgegen dem Gefälle des elektrochemischen Potentials möglich ist, beschränken. Inzwischen hat man aber in der Erythrocytenmembran ein Ferment entdeckt (DUNHAM u. GLYNN[6], POST u. Mitarb.[46]), das wahrscheinlich eine wichtige Rolle beim aktiven Transport spielt. Physikochemische Untersuchungen über den Wirkungsmechanismus dieses Fermentes werden sicher in der nächsten Zukunft für die Permeabilitätsforschung von großem Interesse sein. In den folgenden Abschnitten soll daher der Weg, der zur Entdeckung dieses Fermentes führte, kurz beschrieben werden. Daran anschließend soll die von HOKIN u. HOKIN[26, 27] in einer großen Zahl von Veröffentlichungen vertretene Ansicht diskutiert werden, daß die Phosphatidsäure, die im Intermediärstoffwechsel der Membranphosphatide eine wichtige Stellung einnimmt, beim aktiven Na^+-Transport eine Carrier-Funktion hat.

Die Kinetik des aktiven Transportes

Werden Erythrocyten in Lösungen steigender Kaliumkonzentration suspendiert und wird dann mit K^{42} das Eindringen der Kaliumionen in die Erythrocyten gemessen, so ergibt sich nach SHAW (1954)[52] und GLYNN (1955)[18] folgendes Bild: mit steigender Kaliumkonzentration nimmt die Geschwindigkeit der Kaliumaufnahme zu. Bei hohen K^+-Konzentrationen flacht die Kurve ab und steigt unter geringem Neigungswinkel nahezu proportional der Kaliumkonzentration der Außenlösung weiter an. Dieser Kurvenverlauf läßt sich am einfachsten durch die bereits (s. S. 69) benutzte Annahme erklären, daß die Kaliumaufnahme gleichzeitig sowohl durch aktiven Transport als auch durch passive Ionenwanderung erfolgt. Der erste gekrümmte Abschnitt der Kurve kann durch eine

Formel von der Gestalt der Michaelis-Menten-Gleichung dargestellt werden. Damit wird der Gedanke nahegelegt, daß der Anfangsabschnitt der Kurve die aktive Komponente des K^+-Transportes beschreibt und die Krümmung die Sättigung eines Carriersystems oder die maximale Aktivierung eines am Transport beteiligten Fermentes andeutet. Der lineare Teil der Kurve repräsentiert nach dieser Vorstellung die bei Diffusionsvorgängen zu erwartende Proportionalität zwischen K^+-Konzentration und Ionenfluß. Diese Auffassung erhielt eine Stütze durch die Beobachtung, daß die lineare Komponente des Ionenflusses durch Substratmangel nicht verändert wird und daß die der Michaelis-Konstanten entsprechende Größe vom Stoffwechselzustand der Zellen unabhängig ist. Allein der die Maximalkonzentration der Carriermoleküle repräsentierende Parameter nimmt beim Fehlen von Substrat einen niedrigeren Wert an. Damit war gezeigt, daß es möglich ist, mit Hilfe kinetischer Untersuchungen zwischen aktivem und passivem Anteil des Kationenflusses zu unterscheiden. Von dieser Möglichkeit wurde mit großem Gewinn bei Untersuchungen über die wechselseitigen Beziehungen zwischen Kalium- und Natriumtransport Gebrauch gemacht.

Koppelung zwischen Kalium- und Natriumtransport[23]

Ähnlich wie bei der Kaliumpermeabilität gibt es nach GLYNN [18,20] auch beim Natriumtransport eine passive und eine aktive Komponente. Der passive Natriumionenfluß ist von der Kaliumkonzentration der Außenlösung unabhängig, während der aktive Natriumtransport durch Kaliumionen aktiviert wird. Bei vollständigem Fehlen von Kaliumionen in der Außenlösung kann selbst in Gegenwart von Substrat kein aktiver Natriumtransport erfolgen. Die Aktivierung des Natriumtransportsystems durch extracelluläre K^+-Ionen läßt sich durch die Michaelis-Menten-Gleichung beschreiben. Die Michaelis-Konstante ist dabei ebenso groß wie die Michaelis-Konstante, welche die Beziehung zwischen aktivem Natriumtransport und der Kaliumkonzentration beschreibt. Diese Beobachtung spricht für die Auffassung, daß beim aktiven Transport für jedes Kaliumion, das in die Zelle hineinbefördert wird, ein Natriumion an die Außenlösung abgegeben werden muß. Es ist jedoch gegenwärtig noch nicht entschieden, ob dieses Verhältnis ein für allemal fest eingestellt ist, oder ob es an wechselnde Bedürfnisse der Zelle angepaßt werden kann.

Die Hemmung des aktiven Transportes durch Herzglykoside

Durch die Gegenwart von Herzglykosiden werden die Kationenverschiebungen, die bei längerer Lagerung der Erythrocyten in der Kälte eintreten, (vgl. S. 69) nicht verhindert. Die Glykoside hemmen jedoch die nach Erwärmen und Substratzusatz erfolgende Reakkumulation der Kationen vollständig, ohne daß dabei die Glucolyse oder die Resynthese von ATP vermindert wird, SCHATZMANN (1952)[49], WHITTAM (1958)[65]. Die Bedeutung dieses Befundes war offensichtlich. Man konnte sich durch ein näheres Studium der Herzglykosidwirkung detaillierte Auskunft über die Transportreaktion selbst versprechen. GLYNN[19] führte eine sorgfältige Analyse der durch Herzglykoside hervorgerufenen Änderungen der Kinetik des Kationentransportes durch. Dabei gelangte er zu der Auffassung, daß die Herzglykoside bei Menschenerythrocyten (im Gegensatz z. B. zu Schaferythrocyten, (TOSTESON u. HOFFMAN, 1960)[60] nicht nur den aktiven Transport vollständig blockieren, sondern gleichzeitig die passiven K^+- und Na^+-Bewegungen vermindern. Vergleichende Studien mit verschiedenen Herzglykosiden ergaben[19], daß die Wirksamkeit entscheidend von der Konfiguration des Glykosids am C_{17} und dem Grad der Sättigung des Lactonringes abhängt. Die Hemmwirkung der Herzglykoside kann durch eine Erhöhung der K^+-Konzentration im Suspensionsmedium vermindert oder aufgehoben werden.

ATP als Energielieferant für den aktiven Transport

Es lag nahe anzunehmen, daß der aktive Transport auf Kosten des Abbaues von energiereichem Phosphat vor sich geht. Diese Vermutung konnte mit Hilfe einer ingeniösen, von STRAUB u. Mitarb. (1952)[56a,15] erdachten Versuchsanordnung bestätigt werden. STRAUB gelang es, den Zellinhalt weitgehend aus den Erythrocyten zu entfernen und durch bekannte Substrate zu ersetzen, indem er Erythrocyten in hypotonischen Lösungen der betreffenden Substrate hämolysierte. Im Augenblick der osmotischen Hämolyse werden in der Zellmembran Kanäle geöffnet, durch die der Zellinhalt in die Außenlösung und Substrate ins Zellinnere strömen. Die Membran erleidet dabei keine irreversible Schädigung[57]. Sie gewinnt die Fähigkeit zurück, der Diffusion der Kationen einen hohen Widerstand entgegenzusetzen (TEORELL)[59] und unter Umsatz der im Zellinneren gefangenen Substrate aktiven Ionentransport zu bewerkstelligen.

STRAUB u. Mitarb. konnten mit dieser Technik vor allem zeigen, daß ATP-gefüllte Stromata, die wegen Mangel an geeigneten Substraten nicht glucolysieren können, in Gegenwart von intracellulärem Magnesium imstande sind, unter ATP-Spaltung Kationen entgegen den Gradienten des elektrochemischen Potentials zu akkumulieren[56a,15,47]. HOFFMAN (1960)[3] war dann schließlich in der Lage mit einer wesentlich verbesserten Technik nachzuweisen, daß nur ATP ein für den aktiven Kationentransport der Erythrocytenstromata brauchbares Substrat ist, während die anderen Nucleotide wie UTP, ITP, GTP oder CTP ungeeignet sind.

Die ATP-ase-Aktivität der Erythrocytenmembran

Im Besitz der eben geschilderten Kenntnisse war es naheliegend, nach einem spezifisch mit ATP reagierenden Ferment zu suchen, das in der Erythrocytenmembran lokalisiert ist, durch K^+, Na^+ und Mg^{++} aktiviert und durch Herzglykoside gehemmt wird. Bereits von STRAUB u. Mitarb.[15a,56b] war nachgewiesen worden, daß die Erythrocytenmembran ATP-ase-Aktivität besitzt. Es gelang STRAUB aber nicht, eine Beziehung zwischen ATP-ase-Aktivität von Stroma und aktivem Transport aufzufinden. Kürzlich konnten nun DUNHAM u. GLYNN[6] sowie POST u. Mitarb.[46] in der Erythrocytenmembran eine ATP-ase nachweisen, die alle geforderten Eigenschaften hinsichtlich Spezifität, Aktivierbarkeit und Hemmbarkeit besitzt. Die Eigenschaften dieser „ATP-ase“ und des Transportsystems stimmen selbst in nur quantitativ faßbaren Details überein, und die charakteristischen Unterschiede der Wirksamkeit verschiedener Typen von Herzglykosiden bei der Hemmung des Kationentransportes werden auch bei der Fermenthemmung beobachtet. DUNHAM und GLYNN konnten sogar zeigen, daß eine Erhöhung der K^+-Konzentration eine Verminderung der Glykosidwirkung auf Transport und Ferment herbeiführt. Es ist also sehr wahrscheinlich, daß eine enge Beziehung zwischen ATP-ase-Aktivität der Erythrocytenmembran und aktivem Kationentransport besteht. Es ist aber noch unklar, welcher Art diese Beziehung ist. Einerseits ist es vorstellbar, daß die fermentative Abspaltung von anorganischem Phosphat aus dem ATP-Molekül unmittelbar mit einer Translokation von K^+ und Na^+ verbunden ist. Andererseits muß aber daran gedacht werden, daß die „ATP-ase“ vielleicht in der intakten Zelle eine Transphosphorylierung bewerkstelligt. Bei dem beschriebenen Ferment würde es sich dann um eine Kinase handeln, welche nur einen Schritt aus einer Folge von am aktiven Transport beteiligten Reaktionen katalysiert*.

Die Rolle der Phosphatidsäure

Diese letztere Auffassung erhielt nun kürzlich eine wesentliche Stütze durch Experimente von HOKIN u. HOKIN[28]. Diese Autoren hatten die Beziehung zwischen dem Stoffwechsel der Phosphatide und der Sekretion von organischen Substanzen, wie Eiweißkörpern und Polypeptiden, untersucht und dabei gefunden, daß die sekretorische Aktivität in allen daraufhin überprüften Fällen von einer Vermehrung des Umsatzes an Inositphosphatiden und Phosphatidsäure begleitet ist. Auf Grund dieser Ergebnisse wurden HOKIN u. HOKIN[26] dazu veranlaßt, die Frage aufzuwerfen, ob diese Phosphatide eine Rolle beim aktiven Kationentransport spielen könnten**. Sie benutzten für ihre Versuche die Salzdrüse des Albatros, die nach Stimulierung mit Acetylcholin eine fast 1 molare, praktisch proteinfreie NaCl-Lösung produzieren kann. Unter Acetylcholin vermehrte NaCl-Produktion war regel-

* Es scheint, als ob viele der bei Erythrocyten beobachteten Eigenschaften des Transportsystems für Alkalimetallionen auch in anderen Zellen und Geweben vorkommen[62a]. Insbesondere hat sich herausgestellt, daß in Muskel[28a], Nerv, Niere und Froschhaut[29] der aktive Transport durch Herzglykoside gehemmt werden kann und daß ein Ferment, welches der oben beschriebenen „ATP-ase" ähnlich ist, in Bakterienzellen vorkommt (SOLOMON, persönl. Mitt.). — In diesem Zusammenhang verdient noch folgendes hervorgehoben zu werden: Herzglykoside können den Transport von Hexosen im Darm[4] und von sauren Farbstoffen in Gewebsschnitten aus Rattennieren (BRAUN[2]) hemmen. In letzteren wird die Glykosidwirkung ähnlich wie beim Transport von Alkalimetallionen in roten Blutkörperchen durch einen Überschuß von K^+-Ionen weitgehend wieder aufgehoben. Es ist in beiden Fällen denkbar, daß es sich um eine indirekte Wirkung handelt, die erst nach Stilllegung der Alkalimetallionenpumpe eintritt. Es könnte aber auch sein, daß sämtliche Transportsysteme ihre Energie aus dem Zerfall von ATP gewinnen und daß in allen Fällen dieselbe „Transport-ATP-ase" (CSÁKY u. Mitarb.[4]) beteiligt ist.

** Auf die Möglichkeit einer Beteiligung der Phosphatidsäure an der Regulierung der Ionenpermeabilität von Zellmembranen wurde zuerst von VOGT[64a] hingewiesen. Er gab eine kritische Darstellung der Gründe, die für und gegen eine Carrierfunktion der Phosphatidsäure sprechen und er diskutierte die Frage, ob gewisse pharmakologische Wirkungen der Phosphatidsäure und ähnlicher Verbindungen durch Änderungen des Membranwiderstandes für passive Kationenbewegungen erklärt werden können[64b,64c].

mäßig begleitet von einem vermehrten Umsatz von Phosphatidsäure und – wenn auch in geringem Ausmaß – von Inositphosphatid; diese Beobachtung führte zu der Annahme, daß Phosphatidsäure vielleicht als Carrier am aktiven Natriumtransport beteiligt ist. Diese zunächst etwas willkürlich erscheinende Vermutung konnte aber durch Untersuchungen gestützt werden, die von den gleichen Autoren kürzlich veröffentlicht wurden. Sie fanden

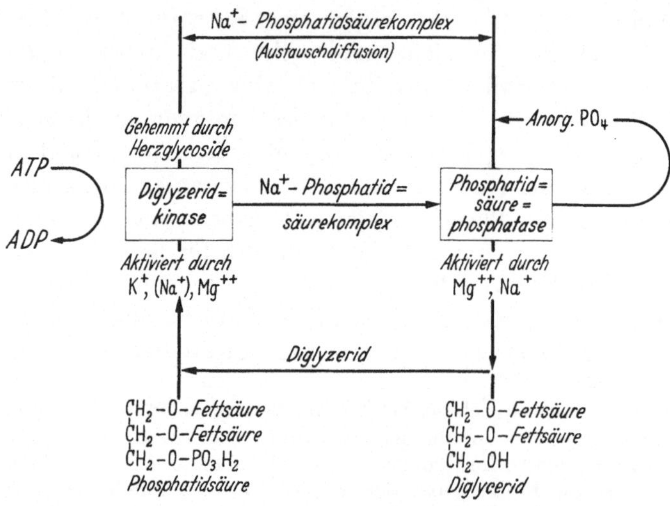

Abb. 6. Erklärung im Text

nämlich, daß die Erythrocytenmembran eine Diglyceridkinase enthält[28]. Diese Kinase ist spezifisch für ATP und wird durch Magnesium- und Kaliumionen aktiviert. Die durch Phosphorylierung von Diglycerid entstehende Phosphatidsäure wird durch eine ebenfalls in der Erythrocytenmembran lokalisierte Phosphatidsäurephosphatase gespalten. Diese Phosphatase enthält zwei Komponenten, von denen eine durch Natrium gehemmt und die andere durch Magnesium- und Natriumionen aktiviert wird. HOKIN u. HOKIN vermuten nun, daß es sich bei der von DUNHAM und GLYNN[6] sowie von POST u. Mitarb.[46] beschriebenen ATP-ase in Wirklichkeit um zwei Fermente handelt, nämlich um die von ihnen entdeckten Fermente Diglyceridkinase und Phosphatidsäurephosphatase.

Bereits Anfang vorigen Jahres hatte HOFFMAN[3] beobachtet, daß P^{32} aus ATP^{32} in die Phosphatidsäure von Erythrocytenmembranen eingebaut wird und daß diese Transphosphorylierung ebenso wie der aktive Transport durch Herzglykoside gehemmt wird.

Faßt man alle diese Beobachtungen zusammen, so scheint auf den ersten Blick die Schlußfolgerung sehr plausibel, daß die Phosphatidsäure als Carrier am aktiven Transport beteiligt ist. HOKIN u. HOKIN haben das in der Abb. 6 dargestellte Schema zur Beschreibung des aktiven Natriumtransportes mit Phosphatidsäure als Carrier angegeben[26,27]. Es scheint mir aber, daß dieses Schema die Verhältnisse beim Erythrocyten nicht wirklich befriedigend beschreibt. Wenn es sich bei der Phosphatidsäure tatsächlich um den Natriumcarrier handelte, so sollte sie lediglich Natrium aus dem Zellinneren in die Außenlösung transportieren können. Man müßte dann annehmen, daß die oben beschriebene Koppelung zwischen aktivem Kalium- und Natriumtransport dadurch zustande kommt, daß entweder das in der äußeren Oberfläche entstehende Diglycerid auf dem Rückweg zur Membraninnenseite Kalium bindet oder daß aus dem Diglycerid ein anderes, bevorzugt Kalium-bindendes Phosphatid gebildet wird. Beide Möglichkeiten sind natürlich offen. Es ist aber kaum anzunehmen, daß das Diglycerid Kationen binden kann, und die Bildung irgend eines anderen Phosphatids aus Diglycerid würde die Beteiligung von CTP (Cytidintriphosphat) am aktiven Transport voraus setzen. CTP wird aber für die Ionenakkumulation nicht benötigt[3]. Es bliebe dann noch die Annahme übrig, daß das Natrium durch Phosphatidsäure als Carrier transportiert und Kalium während der zur Bildung des Natriumcarriers führenden Transphosphorylierungsreaktion akkumuliert wird. Es kann aber noch an eine gänzlich andersartige Möglichkeit gedacht werden.

GLYNN[19] hatte bei Menschenerythrocyten eine Hemmung passiver Kalium- und Natriumbewegungen durch Herzglykoside gefunden. Es wäre durchaus denkbar, daß die Phosphatidsäure ähnlich wie manche Detergentien die passive Penetration durch die Zellmembran begünstigt. Man könnte dementsprechend vermuten, daß die Tätigkeit der Ionenpumpe, die in einer Translokation von Kalium und Natrium während der Transphosphorylierung durch Diglyceridkinase besteht, gleichzeitig die Größe des Lecks, durch das passive Ionenbewegungen erfolgen, mitbestimmt. Diese Frage sollte sich dadurch klären lassen, daß man Phosphatidsäure zu einer Erythrocytensuspension hinzufügt und untersucht, ob Permeabilitätsänderungen eintreten. Falls die Phosphatidsäure lediglich als spezifischer Natriumcarrier wirken sollte, müßte man erwarten, daß eine künstliche Erhöhung ihrer Konzentration in der Zellmembran lediglich zu einer Vergrößerung der bereits von GLYNN beobachteten Austauschdiffusion von Natriumionen führt[18,20]. Mit anderen Worten: Durch ein Überangebot an Carriermolekülen müßten Natriumionen aus dem Zellinneren gegen Natriumionen aus der

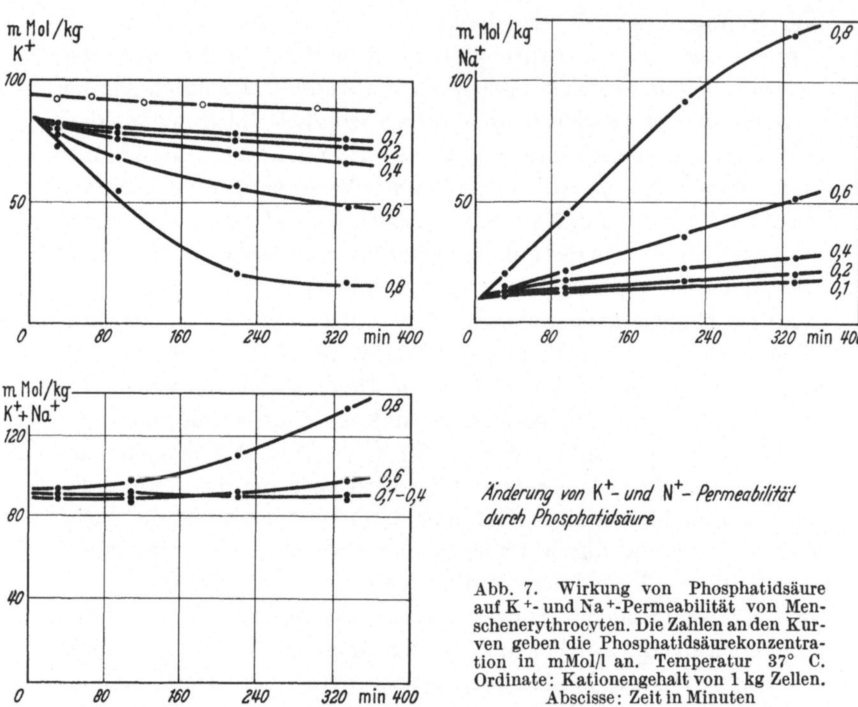

Abb. 7. Wirkung von Phosphatidsäure auf K^+- und Na^+-Permeabilität von Menschenerythrocyten. Die Zahlen an den Kurven geben die Phosphatidsäurekonzentration in mMol/l an. Temperatur 37° C. Ordinate: Kationengehalt von 1 kg Zellen. Abscisse: Zeit in Minuten

Außenlösung ausgetauscht werden, ohne daß es dabei zu nennenswerten Nettoverschiebungen von Natriumionen kommt. Wenn aber die Phosphatidsäure den Membranwiderstand für passive Kalium- und Natriumionenbewegungen vermindern sollte, so müßte man eine mit steigender Phosphatidsäurekonzentration zunehmend rascher erfolgende, bergab gerichtete Penetration von Kalium- und Natriumionen beobachten.

Zur Überprüfung dieser Frage haben wir Versuche über die Wirkung von Phosphatidsäure auf die Kationenpermeabilität von

Menschenerythrocyten ausgeführt. Dazu haben wir uns aus chromatographisch gereinigtem Eierlecithin unter Verwendung einer aus Rosenkohl isolierten Phospholipase reine Phosphatidsäure hergestellt. Die Abb. 7 zeigt zunächst, daß das Eierlecithin, welches als Ausgangssubstanz für die Phosphatidsäureherstellung verwendet wurde, keinen nennenswerten Einfluß auf Kalium- und Natriumpermeabilität ausübt. Phosphatidsäure bewirkt dagegen raschen Kaliumverlust, der von Na^+-Aufnahme begleitet wird. Bei nicht zu hoher Phosphatidsäurekonzentration bleibt die Summe Kalium plus Natrium annähernd konstant. Die Abb. 7 zeigt außerdem, daß Zusatz des die Phosphatidsäurephosphatase aktivierenden Magnesiums die Wirksamkeit der Phosphatidsäure kaum vermindert, während Calcium eine außerordentlich starke Hemmwirkung entfaltet*. Gleichzeitiger Zusatz von Inosit, Cholin, Serin und Colamin beeinflussen die Wirkung der Phosphatidsäure nicht.

Unsere Versuche erlauben es nicht, die Möglichkeit auszuschließen, daß geringe Mengen an Phosphatidsäure eine Erhöhung der Austauschdiffusion für Natriumionen bewirken. Wir können also durch unsere Versuche nichts zur Beantwortung der Frage beitragen, ob es sich bei der Phosphatidsäure wirklich um den Natriumionencarrier handelt. Unsere Experimente gestatten jedoch eine Aussage: Falls es abnorme Stoffwechselbedingungen geben sollte, die zu einer Vergrößerung der Phosphatidsäurekonzentration in der Erythrocytenmembran führen, so sollte man – ebenso wie in unseren Versuchen – eine erhebliche Änderung des Membranwiderstandes für Kalium- und Natriumionen erwarten. Selbst wenn sich also nicht bestätigen sollte, daß Phosphatidsäure eine Carrierfunktion ausübt, so sollte doch damit gerechnet werden, daß sie eine wichtige Rolle bei der Regulierung des Widerstandes spielt, den die Zellmembran der passiven Penetration der Alkalimetallionen bietet.

b) Kontrolle des Membranwiderstandes durch den Zellstoffwechsel. Wir wollen nun die Diskussion des aktiven Transportes abbrechen, dabei aber für die kommenden Betrachtungen den Gedanken aufgreifen, daß der Intermediärstoffwechsel vielleicht den Widerstand einregelt, den die Zellmembran den passiven Ionenbewegungen entgegensetzt[15].

* Im Gegensatz hierzu fördert Calcium die hämolytische Aktivität von Lysolecithin.

Normalerweise repräsentiert die ungleiche Kationenverteilung zwischen Zellinnerem und Serum einen stationären Zustand, wobei aktiver und passiver Transport sich die Waage halten. Die sich im stationären Zustand einstellenden Ionenkonzentrationen hängen dabei von der Förderleistung der Ionenpumpe und dem Widerstand, den die Membran der passiven Penetration entgegensetzt, ab. In einem System, in dem der Membranwiderstand allein vom Porenradius und der Konzentration der Festionen bestimmt wird und in dem Kalium- und Natriumbewegungen starr miteinander gekoppelt sind, so daß für jedes Kaliumion, das in die Zelle hineinbefördert wird, eine bestimmte Anzahl an Natriumionen an die Außenlösung abgegeben werden muß, können Kalium- und Natriumkonzentration nicht getrennt voneinander eingeregelt werden. Das Verhältnis der für beide Ionensorten im stationären Zustand erreichten Konzentrationen wird bei jeder Akkumulationsrate allein vom Verhältnis der Penetrationswiderstände für die passive Permeabilität beider Ionensorten bestimmt. Um die Frage beurteilen zu können, ob die Zelle funktionelle Änderungen im Haushalt der einzelnen Ionensorten vornehmen kann, haben wir herauszufinden versucht, ob in der Erythrocytenmembran Mechanismen vorhanden sind, die eine getrennte Kontrolle von Kalium- und Natriumkonzentration gestatten. Wir haben uns dabei nicht mit der Frage beschäftigt, ob die Koppelung der aktiven K^+- und Na^+-Bewegungen geändert werden kann. Wir haben statt dessen nach Bedingungen gesucht, unter denen der Membranwiderstand, welcher der passiven Penetration der Alkalimetallionen entgegengesetzt wird, selektiv, d. h. entweder überwiegend für K^+ oder überwiegend für Na^+, beeinflußt werden kann.

Ich möchte nun anhand ausgewählter Beispiele zeigen, daß sich – zumindest bei Menschenerythrocyten – der Membranwiderstand für Na^+ und K^+ durch geeignete Membrangifte oder durch Substanzen, die abnorme Stoffwechselreaktionen auslösen, getrennt beeinflussen läßt.

Die Abb. 8 und 9 zeigen, daß eine ganze Reihe sehr verschiedenartiger Zellgifte Kaliumverlust hervorruft, ohne die Natriumpermeabilität in gleichem Ausmaß zu vergrößern. Da der Kaliumausstrom in allen Fällen sehr viel rascher erfolgt, als es bei einfacher Unterbrechung des aktiven Transportes zu erwarten wäre, darf man annehmen, daß der Kaliumausstrom durch eine drastische Senkung des Membranwiderstandes für K^+-Ionen zustande kommt.

Bei der Bleivergiftung[64,44] werden gleichzeitig Kaliumpermeabilität und Zellstoffwechsel gestört[33]. Dabei handelt es sich offenbar um zwei voneinander unabhängige Wirkungen des Giftes. Der ohne Latenzzeit einsetzende K^+-Verlust beginnt immer eher als der ATP-Zerfall, und gelegentlich gelingt es sogar, die Permeabilitätsänderung auszulösen, ohne daß dabei nennenswerter ATP-Abbau erfolgt. Bleivergiftete, nahezu ATP-freie Erythrocytenstromata schrumpfen in NaCl- und schwellen in KCl-Lösung. Ihre K^+-Permeabilität wird also durch das Schwermetall ebenso verändert wie die der intakten Zellen. Diese und andere Versuche über den Einfluß nicht permeierender Komplexbildner auf die Pb-Vergiftung roter Blutkörperchen weisen darauf hin[43,21], daß das Pb die Permeabilitätsänderung durch chemische Reaktionen mit spezifischen, bisher aber noch nicht identifizierten Receptoren in der Erythrocytenmembran selbst bewirkt.

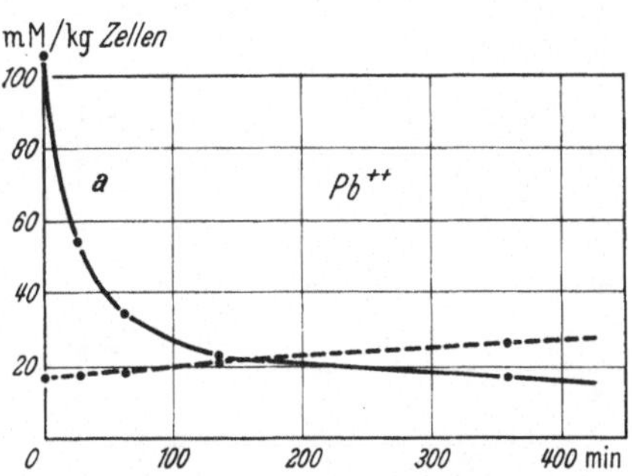

Abb. 8a. K^+-Verlust und Na^+-Aufnahme von Menschenerythrocyten nach Vergiftung mit 0,06 mMol/l $PbCl_2$. Temperatur 21° C. Ausgezogene Kurve: K^+-Gehalt, gestrichelte Kurve: Na^+-Gehalt

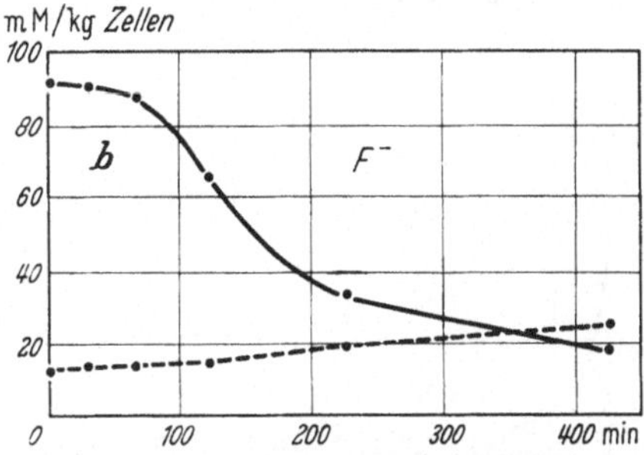

Abb. 8b. K^+-Verlust und Na^+-Aufnahme nach Vergiftung mit 40 mMol/l NaF. Menschenerythrocyten. Temperatur 37° C. Ausgezogene Kurve: K^+-Gehalt, gestrichelte Kurve: Na^+-Gehalt

Die Natur der durch das Glykolysegift (Natriumfluorid) hervorgerufenen Permeabilitätsänderung (WILBRANDT[69,70]) hängt entscheidend von der Konzentration der Erdalkalimetallionen in der äußeren Membranoberfläche ab. In Gegenwart von erdalkalimetallbindenden Komplexbildnern[31,14] hemmt Fluorid lediglich den aktiven Transport. Der Membranwiderstand bleibt unverändert[21]. Bei sehr hohen Konzentrationen von Mg^{++}, Ca^{++} oder Sr^{++} wird der Membranwiderstand für K^+ und Na^+ unmittelbar nach Zusatz des Giftes stark vermindert[32]. Der beobachtete Effekt ähnelt der unspezifischen Wirkung, die Hämolytica ausüben, wenn sie in sublytischen Dosen zu Erythrocyten hinzugefügt werden. Bei geringen Erdalkalimetallkonzentrationen beobachtet man jedoch eine erst nach einer Latenzzeit einsetzende spezifische Verminderung des Membranwiderstandes für Kaliumionen, während die Na^+-Permeabilität nur wenig zunimmt[32]. Detaillierte Untersuchungen über den Mechanismus dieser Fluoridwirkung auf die Kaliumpermeabilität haben folgendes Bild ergeben[34, 35]: Der Kaliumverlust wird — wenn auch nicht ausschließlich — durch die Bildung eines Erdalkalimetall-Fluorid-Komplexes mit Liganden in der äußeren Zelloberfläche ausgelöst. In der intakten Zellmembran kann dieser Komplex jedoch nicht entstehen. Die durch das Fluorid hervorgerufene Glucolysehemmung bewirkt aber während der dem Kaliumverlust vorangehenden Latenzzeit ATP-Zerfall und andere Störungen des intermediären Stoffwechsels. Dadurch wird die Zellmembran aus noch ungeklärten Gründen so verändert, daß nun die Bildung eines Erdalkalimetall-Fluorid-Komplexes mit Receptoren in der Zelloberfläche erfolgen kann und der Kaliumverlust einsetzt.

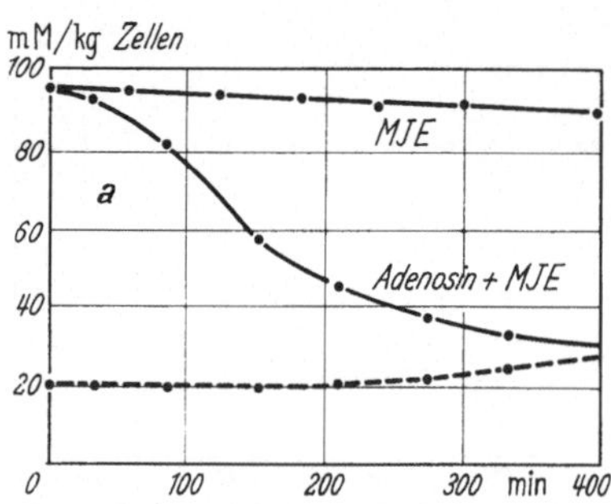

Abb. 9a. K^+-Verlust und Na^+-Aufnahme nach Vergiftung mit 2,5 mMol/l Monojodessigsäure (MJE) und 2,5 mMol/l Monojodessigsäure + 10 mMol/l Adenosin (MJE + Adenosin). $CaCl_2$-Konzentration 0,5 mMol/l. Menschenerythrocyten. Temperatur 37° C. Ausgezogene Kurve: K^+-Gehalt, gestrichelte Kurve: Na^+-Gehalt

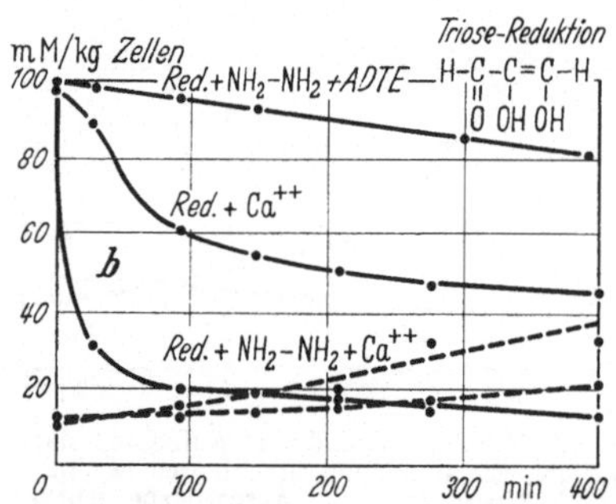

Abb. 9b. K^+-Verlust und Na^+-Aufnahme nach Vergiftung mit 2,5 mMol/l Triosereducton + 0,5 mMol/l $CaCl_2$ (Red. + Ca^{++}). 2,5 mMol/l Triosereducton + 0,5 mMol/l $CaCl_2$ + 10 mMol/l Hydrazin (Red. + Ca^{++} + NH_2NH_2). 2,5 mMol/l Triosereducton + 10 mMol/l NH_2NH_2 + 10 mMol/l Äthylendiamintetraessigsäure (Red. + NH_2NH_2 + ADTE). Menschenerythrocyten. Temperatur 37° C. Ausgezogene Kurve: K^+-Gehalt, gestrichelte Kurve: Na^+-Gehalt

In Gegenwart von Monojodacetat (MJE) erfolgt eine vollständige Hemmung des aktiven Transportes. Diese Hemmung beruht lediglich auf dem durch das Stoffwechselgift herbeigeführten ATP-Zerfall, und weder die Ionenpumpe noch der Membranwiderstand werden durch das Gift unmittelbar beeinflußt (HOFFMAN[3]). Fügt man nun zu MJE-vergifteten Menschenerythrocyten Adenosin hinzu, so erfolgt nach einer gewissen Latenzzeit die in Abb. 9a dargestellte spezifische Senkung des Membranwiderstandes für Kaliumionen (GARDOS[12, 13, 14]). Die Permeabilitätsänderung tritt aber nur ein, wenn Ca^{++}-Ionen vorhanden sind, die in diesem Falle auch nicht durch Magnesium ersetzt werden können. Trotz der Gegenwart des Stoffwechselgiftes wird das Adenosin von den Zellen umgesetzt. Es wird zunächst desaminiert, und das dabei entstehende Inosin wird anschließend phosphorolytisch gespalten. Die Permeabilitätsänderung kann nur eintreten, wenn das entstehende Pentosephosphat umgesetzt wird. Die Zellen synthetisieren daraus ein Stoffwechselprodukt, das normalerweise nicht oder nur in geringem Ausmaß gebildet wird und das offenbar in Reaktionen des Membranstoffwechsels eingreift, die für die Aufrechterhaltung des normalen Membranwiderstandes für K^+-Ionen von entscheidender Bedeutung sind.

Wir haben eine Reihe von Versuchen ausgeführt*, um einen Anhaltspunkt dafür zu gewinnen, welcher Seitenweg des Zellstoffwechsels diesen in sehr geringer Konzentration spezifisch auf die Kaliumpermeabilität wirkenden Stoff bildet. Zu unserer Überraschung ergab sich, daß Inosin sehr viel weniger Kaliumverlust bewirkt als Adenosin. Setzt man aber zum Inosin steigende Mengen an NH_4 hinzu, so wird die Wirksamkeit des Inosins verstärkt, bis schließlich — in Gegenwart eines großen Überschusses an NH_4Cl — die durch Inosin und Adenosin hervorgerufenen Effekte sich nicht mehr voneinander unterscheiden lassen (Abb. 10).

Wenn Hydrazin anstelle von Amoniumionen zusammen mit Adenosin oder Inosin zu MJE-vergifteten Zellen hinzugesetzt wird, so bewirken beide Substrate gleichgroße Änderungen der K-Permeabilität, wobei die Permeabilitätsänderung nach kürzerer Latenzzeit einsetzt als in Gegenwart von NH_4Cl. Diese Befunde und weitere, mehr ins Detail gehende Untersuchungen haben uns

---

* Die Untersuchungen wurden gemeinsam mit Frl. S. LEPKE ausgeführt.

zu der Auffassung gebracht, daß das durch den Umsatz der Nucleoside entstehende Pentosephosphat in Gegenwart von MJE eine abnorme, d. h. nicht in den Pentosephosphatcyclus einmündende Umsetzung erleidet und daß eines der entstehenden Produkte mit NH_3 oder NH_2—NH_2 eine stickstoffhaltige Verbindung bildet, die dann unter (vielleicht katalytischer) Beteiligung von Ca^{++} in die den Kaliumverlust auslösende Substanz übergeht.

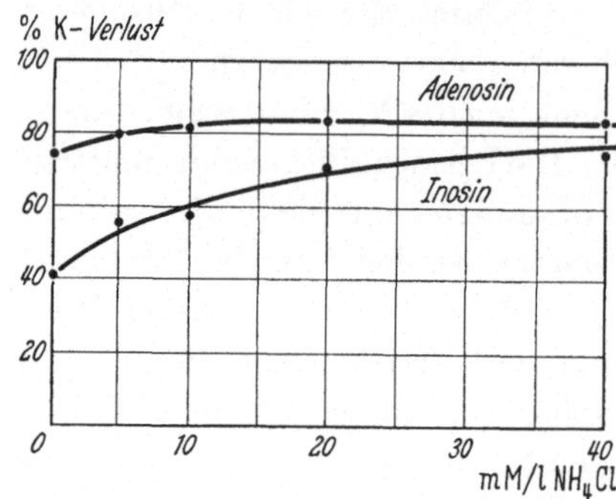

Abb. 10. K^+-Verlust in Gegenwart von 2,5 mMol/l Monojodacetat und Adenosin oder Inosin (10 mMol/l) sowie steigenden Zusätzen an NH_4Cl. Ca^{++}-Konzentration der Außenlösung 0,5 mMol/l. Menschenerythrocyten. Temperatur 37° C. Ordinate: Kaliumverlust nach 380 min. Abscisse: NH_4Cl-Konzentration im Suspensionsmedium

Auf der Suche nach dem aus dem Pentosephosphat entstehenden abnormen Stoffwechselprodukt prüften wir die Wirkung von Triosereducton* auf die Erythrocytenpermeabilität und fanden dabei eine weitere spezifisch die Kaliumpermeabilität beeinflussende Substanz (Abbildung 9b). Diese reaktionsfähige Verbindung bewirkt in Gegenwart von Ca^{++} — bei nur geringer Na^+-Aufnahme — raschen Kaliumverlust, der durch NH_2—NH_2 erheblich verstärkt, durch NH_4Cl aber gehemmt wird.

Permeabilität für Erdalkalimetallionen[44b]

Solange der Zellstoffwechsel normal abläuft, ist die Membran von Menschenerythrocyten für Mg^{++}, Ca^{++} und Sr^{++} nahezu undurchlässig (Abb. 11). Dabei ist es gleichgültig, ob Glucose, Inosin oder Adenosin als Substrate umgesetzt werden. Wenn jedoch Mangel an Substrat eintritt oder wenn in Gegenwart von Glucose als Substrat die Glucolyse durch Monojodessigsäure gehemmt wird, so dringen Ca^{++} und Sr^{++} relativ rasch ins Zellinnere ein. Die Verteilung von Ca^{++} und Sr^{++} zwischen Zellen und Suspensionsmedium (isot. NaCl) wird also offensichtlich durch den Zellstoffwechsel kontrolliert. Trotz zahlreicher Bemühungen ist

* Wir danken Herrn Dr. SCHNIEWIND für die Überlassung von schmelzpunktreinem Triosereducton.

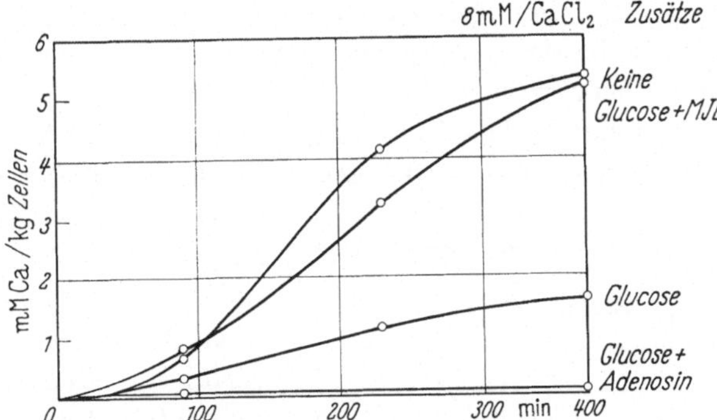

Abb. 11. Wirkung von Substraten und Stoffwechselgiften auf die Durchlässigkeit der Zellmembran von Menschenerythrocyten für Calciumionen. Gewaschene Zellen in isotonischer NaCl-Lösung suspendiert. Suspensionsdichte 40 Vol.-%. $CaCl_2$-Konzentration zu Versuchsbeginn im Medium: 8 mMol/l. Temperatur 37° C. Ordinate: Ca^{++}-Gehalt der Zellen in mMol/kg. Abszisse: Zeit in Minuten

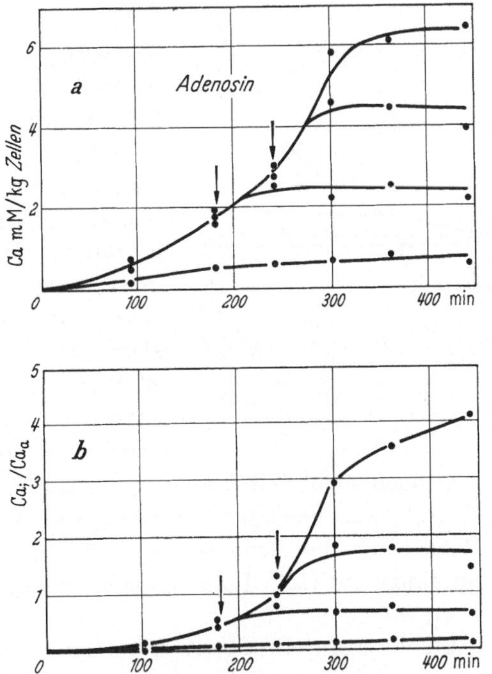

Abb. 12a u. b. Hemmung der Ca^{++}-Aufnahme aus substratfreier Kochsalzlösung durch Zusatz von Adenosin. Die Pfeile geben den Zeitpunkt des Substratzusatzes an. Suspensionsdichte 40 Vol.-%. Temperatur 37° C. Ordinate: a) Ca^{++}-Gehalt der Zellen in mMol/kg. b) Ca^{++}-Verteilung zwischen Zellwasser und Außenlösung. Abszisse: Zeit in Minuten

es uns aber nicht gelungen, einmal eingedrungenes Ca^{++} durch nachträglichen Zusatz von geeigneten Substraten wieder aus den Erythrocyten herauszubefördern (Abb. 12). Wir vermuten daher, daß — falls es eine Erdalkalimetallionenpumpe in den Menschenerythrocyten geben sollte — die Leistungsfähigkeit dieser Pumpe recht gering ist. Die ungleiche Verteilung der Erdalkalimetallionen zwischen Zellen und Medium dürfte vor allem dadurch aufrechterhalten werden, daß die Zellmembran bei normal ablaufendem Stoffwechsel der passiven Penetration von Mg^{++}, Ca^{++} und Sr^{++} einen sehr hohen Widerstand entgegensetzt (d. h., daß sie weitgehend impermeabel ist). Diese Undurchlässigkeit der Zellmembran ist nicht an die Gegenwart von intracellulärem ATP gebunden. So gelingt es beispielsweise, durch Zusatz von Monojodessigsäure das intracelluläre ATP vollständig zum Verschwinden zu bringen. Dennoch wird die Zellmembran für Ca-Ionen nicht durchlässig, falls nur Adenosin oder Inosin als Substrate zur Verfügung stehen (Abb. 13). Offenbar reichen die Reaktionen des Pentosephosphatcyclus aus, um die Zellmembran in einem für Ca^{++}-Ionen unpassierbaren Zustand zu erhalten.

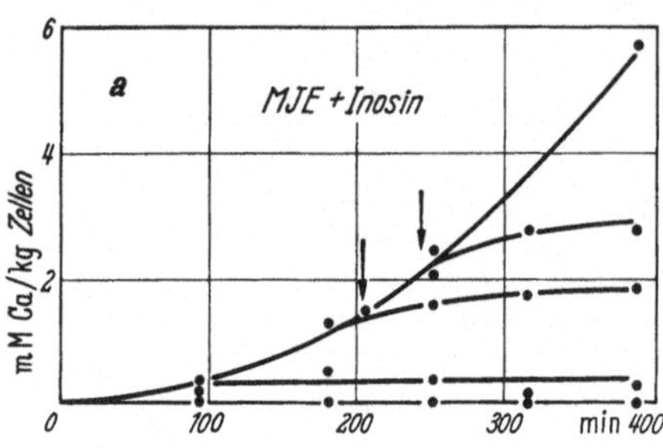

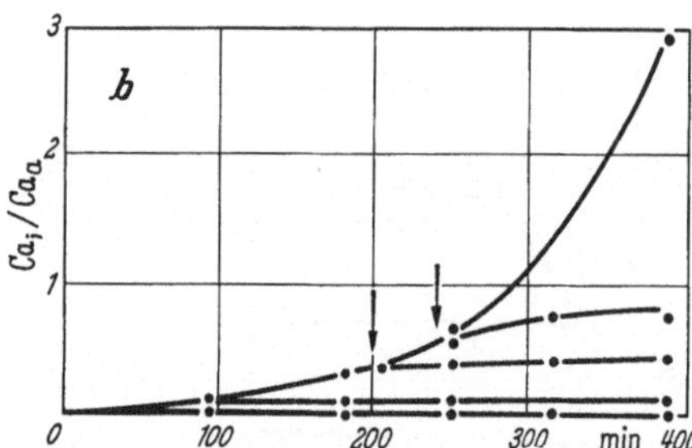

Abb. 13. Hemmung der Ca^{++}-Aufnahme aus substratfreier Kochsalzlösung durch Zusatz von Inosin zu mit Monojodacetat vergifteten Zellen. Die Pfeile geben den Zeitpunkt des Substratzusatzes an. Suspensionsdichte 40 Vol.-%. Temperatur: 37° C. Ordinate: a) Ca^{++}-Gehalt der Zellen in mMol/kg, b) Ca^{++}-Verteilung zwischen Zellwasser und Außenlösung. Abszisse: Zeit in Minuten

Die Ca^{++}-Aufnahme durch substratfrei incubierte Zellen beginnt erst nach einer von der Calciumkonzentration abhängigen Latenzzeit (Abb. 14). Wir glauben daher, daß die Unterbrechung des Zellstoffwechsels nur *eine* Voraussetzung für das Eindringen des

Calciums in die Zellen darstellt. Das Calcium selbst (ebenso wie Strontium) scheint mit konzentrationsabhängiger Geschwindigkeit in der Zellmembran zusätzliche Veränderungen hervorzurufen,

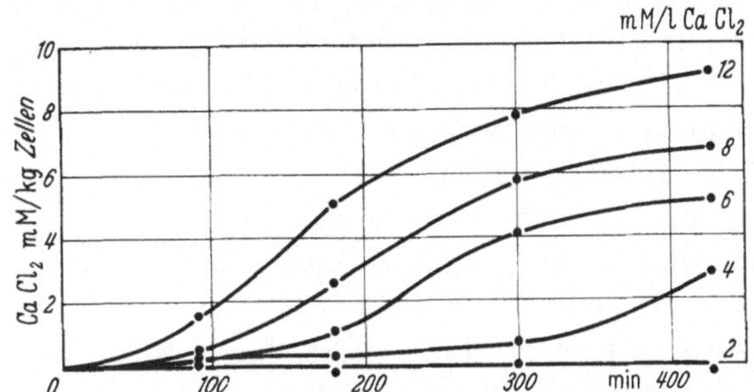

Abb. 14. Ca^{++}-Aufnahme durch gewaschene Menschenerythrocyten aus isotonischer Kochsalzlösung bei verschiedenen Ca^{++}-Konzentrationen. Suspensionsdichte: 40 Vol.-%. Temperatur: 37° C. Ordinate: Ca^{++}-Gehalt der Zellen in mMol/kg. Abszisse: Zeit in Minuten

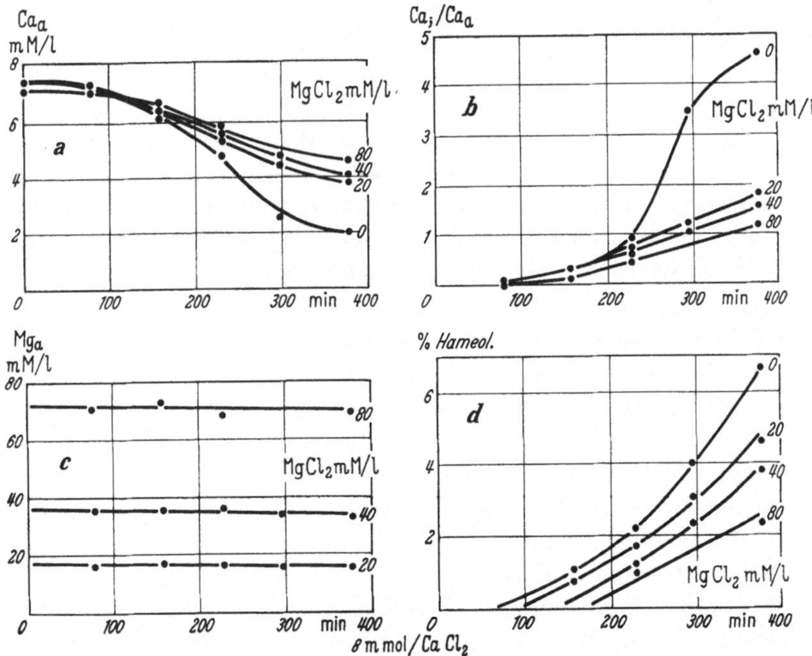

Abb. 15. a) Hemmung der Ca^{++}-Aufnahme aus substratfreier Kochsalzlösung durch steigende $MgCl_2$-Konzentrationen. $A = 0$ mMol/l, $B = 18$ mMol/l, $C = 37$ mMol/l, $D = 71$ mMol/l. b) Ca^{++}-Verteilung zwischen Zellwasser und Außenlösung. c) Mg-Aufnahme aus der Außenlösung. d) Hämolyse (in % der gesamten Zellmenge). Suspensionsdichte 40 Vol.-%. Temperatur: 37° C. Abszisse: Zeit in Minuten

durch die der Weg ins Zellinnere freigemacht wird. Man könnte z. B. daran denken, daß Ca^{++} eine Phospholipase aktiviert und damit den Abbau von Phosphatiden auf abnorme Bahnen lenkt.

Die Ca^{++}(und Sr^{++})-Aufnahme durch die Erythrocyten kann durch das nur sehr langsam penetrierende Magnesium stark gehemmt werden (Abb. 15). Wir halten es für unwahrscheinlich, daß dabei Konkurrenz um ein gemeinsames Transportsystem eine Rolle spielt. Wir glauben statt dessen, daß eine kompetitive Hemmung eines Fermentes in der Zellmembran erfolgt, das nach Aktivierung durch Ca^{++} den Diffusionswiderstand der Membran durch Abbauvorgänge vermindert.

Schließlich ist noch erwähnenswert, daß Kalium- und Calciumionen auf verschiedenen Wegen durch die Membran penetrieren. So wird in Gegenwart von Adenosin und Monojodessigsäure (vgl. S. 85) eine hochgradige Senkung des Membranwiderstandes für K^{+}-Ionen herbeigeführt. Dennoch bleibt die Zellmembran für Ca^{++} unter diesen Versuchsbedingungen nahezu undurchlässig.

Zusammenfassung

Es fragt sich nun, wie sich diese Beobachtungen über die Spezifität der Wirkungen von Membrangiften und abnormen Stoffwechselprodukten auf die passive Kationenpermeabilität in das eingangs skizzierte Bild von der elektrostatisch bedingten Selektivität der Zellmembran einfügen lassen.

Es wurde hervorgehoben, daß die Anionenpermeabilität (von Pferdeerythrocyten) durch Stoffwechselgifte nicht beeinflußt werden kann. Die Selektivität der Membran wird also im wesentlichen von Strukturelementen in der Zellmembran bestimmt, die sich nach Störungen des Intermediärstoffwechsels nicht oder nur sehr langsam ändern. Es wurde ferner die Ansicht vertreten, daß der Membranwiderstand für die Elektrolyte entscheidend von dem durch die Festionenkonzentration vorgeschriebenen Verteilungskoeffizienten und von dem durch die Porenweite bestimmten Grad der räumlichen Behinderung der diffundierenden Partikel abhängt. Damit sollte aber nicht gesagt werden, daß der Diffusionswiderstand eine ein für allemal während des Reifungsprozesses der Zelle fest eingestellte Größe ist, die von Schwankungen des Funktionszustandes der ausgewachsenen Zelle völlig unabhängig ist.

Die positiv geladenen Poren dürften für Kationen, insbesondere für die zweiwertigen Erdalkalimetallionen, nahezu impermeabel sein. Selbst ein geringer, auf anderem Wege — etwa durch die Lipoidphase — erfolgender Ionenfluß könnte daher quantitativ von ausschlaggebender Bedeutung für die passive Kationenpermeabilität sein. Da die Lipoide der Zellmembran ständigem Stoffwechsel unterliegen[1], kann man sich leicht vorstellen, daß Änderungen des Zellstoffwechsels die Wirksamkeit der Permeabilitätsbarriere in spezifischer Weise modifizieren können.

Insgesamt kann also der Zellstoffwechsel in zweierlei Weise in die Kationenpermeabilität eingreifen.

1. durch eine Beeinflussung des aktiven Ionentransportes und
2. durch Änderung des Membranwiderstandes für die passive Diffusion.

Es muß aber noch offenbleiben, wieweit der letztere Mechanismus unter *physiologischen* Bedingungen an der Regulierung der normalen Verteilung der Alkalimetallionen beteiligt ist.

Literatur

1 ALTMAN, K. J., R. N. WHATMAN and A. K. SOLOMON: The incorporation of C^{14}-acetate into the stroma of the erythrocyte. Arch. Biochem. **33**, 168 (1951).

2 BRAUN, W.: Wirkung von Herzglycosiden auf den aktiven Transport saurer Farbstoffe in Gewebsschnitten aus Rattenniere. Persönl. Mitteilung.

2a BRANS, F. H., E. NOLTMANN u. E. VALHAUS: Über den Stoffwechsel von Ribose-5-Phosphat in Haemolysaten. Biochem. Z. **330**, 483 (1958).

3 *Ciba Foundation Study Group No. 5:* Regulation of the inorganic ion content of cells. London: Churchill 1960.

4 CSÁKY, T. Z., H. G. HARTZOG III and G. W. FERNALD: Effect of digitalis on active intestinal sugar transport. Amer. J. Physiol. **200**, 459 (1961).

5 DAVSON, H., and J. F. DANIELLI: The permeability of natural membranes. Oxford 1943.

6 DUNHAM, E. T., and I. M. GLYNN: Adenosinetriphosphatase activity and the active movements of alkali metal ions. J. Physiol. (Lond.) **156**, 274 (1961).

7 DUNKER, E., u. H. PASSOW: Zwei Arten des Anionenaustausches bei den roten Blutkörperchen verschiedener Säugetiere. Pflügers Arch. ges. Physiol. **256**, 446 (1953).

8 DZIADZKA, H., H. PASSOW u. CHR. WEISS: Untersuchungen über die langsame Anionenpermeabilität roter Blutkörperchen im Gleichgewicht. Ber. ges. Physiol. **172**, 1—3, 102 (1954).

[9] ECKEL, R. E.: Potassium exchange in human erythrocytes. I, II. J. cell. comp. Physiol. **51**, 81, 109 (1958).
[10] FLYNN, F., and M. MAIZELS: Cation control in human erythrocytes. J. Physiol. (Lond.) **110**, 301 (1949).
[11] GABRIO, B. W., and F. M. HUENNEKENS: The role of nucleoside phosphorylase in erythrocyte preservation. Biochim. biophys. Acta **18**, 585 (1955).
[12] GÁRDOS, G.: The role of calcium in the potassium permeability of human erythrocytes. Acta physiol. Acad. Sci. hung. **15**, 2 (1959).
[13] GÁRDOS, G.: The function of calcium in the potassium permeability of human erythrocytes. Biochim. biophys. Acta **30**, 653 (1958).
[14] GÁRDOS, G.: Effect of ethylenediamine-tetraacetate on the permeability of human erythrocytes. Acta physiol. Acad. Sci. hung. **14**, 1 (1958).
[15] GÁRDOS, G., u. F. B. STRAUB: Über die Rolle der Adenosintriphosphorsäure (ATP) in der K-Permeabilität der menschlichen roten Blutkörperchen. Acta physiol. Acad. Sci. hung. **12**, 1 (1956).
[15a] GARZÓ, T., A. ULLMANN u. F. B. STRAUB: Die Adenosintriphosphatase der roten Blutkörperchen. Acta physiol. Acad. Sci. hung. **3**, 513 (1952).
[16] GIEBEL, O., u. H. PASSOW: Die Permeabilität der Erythrozytenmembran für organische Anionen. Pflügers Arch. ges. Physiol. **271**, 378 (1960).
[17] GLASSTONE, S., K. J. LAIDLER and H. EYRING: The theory of rate processes. New York-London: McGraw-Hill 1941.
[18] GLYNN, I. M.: Sodium and potassium movements in human red cells. J. Physiol. (Lond.) **134**, 278 (1956).
[19] GLYNN, I. M.: The action of cardiac glycosides on sodium and potassium movements in human red cells. J. Physiol. (Lond.) **136**, 148 (1957).
[20] GLYNN, I. M.: The ionic permeability of the red cell membrane. Progr. Biophys. Chem. **8**, 242 (1957).
[21] GRIGARZIK, H., u. H. PASSOW: Versuche zum Mechanismus der Bleiwirkung auf die Kaliumpermeabilität roter Blutkörperchen. Pflügers Arch. ges. Physiol. **267**, 73 (1958).
[22] HARRIS, J. E.: The influence of the metabolism of human erythrocytes on their potassium content. J. biol. Chem. **141**, 579 (1941).
[23] HARRIS, J. E.: The linkage of sodium- and potassium-active transport in human erythrocytes. Symp. Soc. exp. Biol. **8**, 228 (1954).
[24] HODGKIN, A. L., and R. D. KEYNES: The potassium permeability of a giant nerve fibre. J. Physiol. (Lond.) **128**, 61 (1955).
[24a] HODGKIN, A. L., and B. KATZ: The effect of sodium ions on the electrical activity of the giant axon of the squid. J. Physiol. (Lond.) **108**, 37 (1949).
[25] HOFFMAN, J. F.: Physiological characteristics of human red blood cell ghosts. J. gen. Physiol. **42**, 9 (1959).
[26] HOKIN, L. E., and M. R. HOKIN: Evidence for phosphatidic acid as the sodium carrier. Nature (Lond.) **184**, 1068 (1959).
[27] HOKIN, L. E., and M. R. HOKIN: Studies on the carrier function of phosphatidic acid in sodium transport. J. gen. Physiol. **44**, 61 (1960).
[28] HOKIN, L. E., and M. R. HOKIN: Diglyceride kinase and phosphatidic acid phosphatase in erythrocyte membranes. Nature (Lond.) **189**, 836 (1961).

28a Johnson, J. A.: Influence of ouabain, strophantidin and dihydrostrophantidin on sodium and potassium transport in frog sartorii. Amer. J. Physiol. **187**, 328 (1956).

29 Koefoed-Johnson, V.: The effect of g-strophantin (ouabain) on the active transport of sodium through the isolated frog skin. Acta physiol. scand. 42, Suppl. **145**, 87 (1957).

30 Lefevre, P. G.: The evidence for active transport of monosaccharides across the red cell membrane. Symp. Soc. exp. Biol. 8, Active Transport and Secretion, 1954.

31 Lepke, S., u. H. Passow: Hemmung des Kaliumverlustes fluoridvergifteter Erythrozyten durch Komplexbildner. Pflügers Arch. ges. Physiol. **271**, 389 (1960).

32 Lepke, S., u. H. Passow: Die Wirkung von Erdalkalimetallionen auf die Kationenpermeabilität fluoridvergifteter Erythrozyten. Pflügers Arch. ges. Physiol. **271**, 473 (1960).

33 Lindemann, B., u. H. Passow: Kaliumverlust und ATP-Zerfall in bleivergifteten Menschenerythrozyten. Pflügers Arch. ges. Physiol. **271**, 369 (1960).

34 Lindemann, B., u. H. Passow: Versuche zur Aufklärung der Beziehung zwischen Glukolysehemmung und Kaliumverlust bei der Fluoridvergiftung von Menschenerythrozyten. Pflügers Arch. ges. Physiol. **271**, 497 (1960).

35 Lindemann, B., u. H. Passow: Die Wirkung von Erdalkalimetallionen auf die Hemmung des Kaliumverlustes fluoridvergifteter Erythrozyten durch Substrate und Stoffwechselgifte. Pflügers Arch. ges. Physiol. **271**, 488 (1960).

36 Lundsgaard-Hansen, P.: Vergleich der Wirkungen von Calcium und Digitalis auf die aktiven und passiven Kaliumbewegungen durch die Erythrozytenmembran. Naunyn-Schmiedeberg's Arch. exp. Path. Pharmakol. **231**, 577 (1957).

37 Maizels, M.: Factors in the active transport of cations. J. Physiol. (Lond.) **112**, 59 (1951).

38 Mond, R.: Umkehr der Anionenpermeabilität der roten Blutkörperchen in eine selektive Durchlässigkeit für Kationen. Pflügers Arch. ges. Physiol. **217**, 5/6, 618 (1927).

39 Netter, H.: Theoretische Biochemie. Berlin-Göttingen-Heidelberg: Springer 1959.

40 Passow, H.: Sulfationenpermeabilität roter Blutkörperchen im Donnan-Gleichgewicht. Abstracts of Communications S. 216, XXth Int. Physiol. Congr., Brüssel (1956).

41 Passow, H., u. S. Lepke: Die Rolle des Magnesiums beim Kaliumverlust fluoridvergifteter Menschenerythrozyten. Pflügers Arch. ges. Physiol. **270**, 63 (1959).

42 Passow, H., H. Lohmann u. V. Privat: Die Abhängigkeit der langsamen Anionenpermeabilität roter Blutkörperchen von der Wasserstoffionenkonzentration. Pflügers Arch. ges. Physiol. **272**, 40 (1960).

43 Passow, H., A. Rothstein and T. Clarkson: The general pharmacology of heavy metals. Pharmacol. Rev. **13**, 185 (1961).

[44] PASSOW, H., u. K. TILLMANN: Untersuchungen über den Kaliumverlust bleivergifteter Menschenerythrozyten. Pflügers Arch. ges. Physiol. **262**, 23 (1955).

[44a] PASSOW, H., and U. WILDE: Metabolic control of calcium permeability in human red cells. Intern. Biophysics Congr. Abstracts of contributed papers 175 (1961).

[45] PONDER, E.: Hemolysis and related phenomena. New York: Grune & Stratton 1948.

[46] POST, R. L., C. R. MERRIT, C. R. KINSOLVING and C. D. ALBRIGHT: Membrane adenosinetriphosphatase as a participant in the active transport of sodium and potassium in human erythrocyte. J. biol. chem. **235**, 1796 (1960).

[47] PRAGAY, D.: Über die Kaliumakkumulation „reversibel hämolysierter" menschlicher Blutkörperchen. Acta physiol. Acad. Sci. hung. **11**, 9 (1956).

[48] RAKER, J. W., I. M. TAYLOR, J. M. WELLER and A. B. HASTINGS: Rate of potassium exchange of the human erythrocyte. J. gen. Physiol. **33**, 691 (1950).

[49] SCHATZMANN, H.-J.: Herzglykoside als Hemmstoffe für den aktiven Kalium- und Natriumtransport durch die Erythrozytenmembran. Helv. physiol. pharmacol. Acta **11**, 346 (1953).

[50] SCHLÖGL, R.: Elektrodiffusion in freier Lösung und geladenen Membranen. Z. physik. Chem., N. F. **1**, H. 5/6, 305 (1954).

[51] SCHWIETZER, C. H., u. H. PASSOW: Kinetik und Gleichgewichte bei der langsamen Anionenpermeabilität roter Blutkörperchen. Pflügers Arch. ges. Physiol. **256**, 419 (1953).

[52] SHAW, T. I.: Potassium movements in washed erythrocytes. J. Physiol. (Lond.) **129**, 464 (1955).

[53] SHAPPARD, C. W., and W. R. MARTIN: Cation exchange between cells and plasma of mammalian blood. J. gen. Physiol. **33**, 703 (1950).

[54] SOLDANO, B. A., and G. E. BOYD: Self-diffusion of anions in strong-base anion exchangers. J. Amer. chem. Soc. **75**, 6099 (1954).

[55] SOLOMON, A. K.: The permeability of the human erythrocyte to sodium and potassium. J. gen. Physiol. **36**, 1, 57 (1952).

[56] SOLOMON, A. K.: Red cell membrane structure and ion transport. J. gen. Physiol. **43**, 5, 1 (1960).

[56a] STRAUB, F. B.: Über die Akkumulation der Kaliumionen durch menschliche Blutkörperchen. Acta physiol. Acad. Sci. hung. **4**, 235 (1953).

[56b] SZEKELY, M., S. MANYAI u. F. B. STRAUB: Die Wirkung der Hämolyse auf den Stoffwechsel der roten Blutkörperchen beim Menschen. Acta physiol. Acad. Sci. hung. **4**, 31 (1953).

[57] SZÉKELY, M., S. MÁNYAI u. F. B. STRAUB: Über den Mechanismus der osmotischen Hämolyse. Acta physiol. Acad. Sci. hung. **3**, 571 (1952).

[58] TEORELL, T.: Zur quantitativen Behandlung der Membranpermeabilität. Z. Elektrochem. **55**, 6, 460 (1951).

[5) TEORELL, T.: Permeability properties of erythrocytes ghosts. J. gen. Physiol. **35**, No. 5, pp 669 (1952).

60 TOSTESON, D. C., and J. F. HOFFMAN: Regulation of cell volume by active transport in high and low potassium sheep red cells. J. gen. Physiol. **44**, 1, 169 (1960).
61 TOSTESON, D. C., and J. S. ROBERTSON: Potassium transport in duck red cells. J. cell. comp. Physiol. **47**, 147 (1956).
62 USSING, H. H.: Transport of ions across cellular membranes. Physiol. Rev. **29**, 127 (1949).
62a USSING, H. H.: The alkali metal ions in isolated systems and tissues. In: The alkali metal ions in biology. Berlin, Göttingen, Heidelberg: Springer-Verlag 1960.
63 VILLEGAS, R., T. C. BARTON and A. K. SOLOMON: The entrance of water into beef and dog red cells. J. gen. Physiol. **42**, 2, 355 (1958).
64 VINCENT, P. C.: The effects of heavy metal ions on the human erythrocyte. II. The effects of lead and mercury. Aust. J. exp. Biol. med. Sci. **36**, 589 (1958).
64a VOGT, W.: The chemical nature of Darmstoff. J. Physiol. (Lond.) **137**, 154—167 (1957).
64b VOGT, W.: Chemische und pharmakologische Eigenschaften saurer Phospholipide (Darmstoff). Arzneimittel-Forsch. **8**, 253 (1958).
64c VOGT, W.: Naturally occurring lipid soluble acids of pharmacological interest. Pharmacol. Rev. **10**, 407 (1958).
65 WHITTAM, R.: Potassium movements and ATP in human red cells. J. Physiol. (Lond.) **140**, 479 (1958).
66 WILBRANDT, W.: Transportsysteme für Zucker. Moderne Probleme der Pädiatrie, vol. 4, p. 30. Basel: Karger 1959.
67 WILBRANDT, W.: Die Kinetik des Ionenaustausches durch selektiv ionenpermeable Membranen. Pflügers Arch. ges. Physiol. **246**, 274 (1942)
68 WILBRANDT, W.: Untersuchungen über langsamen Anionenaustausch durch die Erythrozytenmembran. Pflügers Arch. ges. Physiol. **246**, 291 (1943).
69 WILBRANDT, W.: A relation between the permeability of the red cell and its metabolism. Trans. Faraday Soc. **33**, 959 (1937).
70 WILBRANDT, W.: Die Abhängigkeit der Ionenpermeabilität der Erythrozyten vom glykolytischen Stoffwechsel. Pflügers Arch. ges. Physiol. **243**, 519 (1940).
71 WILBRANDT, W.: Die Ionenpermeabilität der Erythrocyten in Nichtleiterlösungen. Pflügers Arch. ges. Physiol. **243**, 537 (1940).

Diskussion

Diskussionsleiter: DIRSCHERL, *Bonn*

WILBRANDT (Bern): Ich möchte nur eine kurze Bemerkung über den Angriffsort der Herzglykoside bei der Hemmung des aktiven Transportes anfügen. Im Gegensatz zu dem im Vortrag gezeigten Schema dürfte das Herzglykosid nicht auf der Membraninnen-, sondern auf der Membranaußenseite seine Wirkung entfalten. Beim Erythrocyten ist das höchst wahrscheinlich. Bei der Nervenzelle wurde es von CALDWELL nachgewiesen. Es fragt sich nun, wie eine Hemmung der auf der Membraninnenseite gelegenen „ATP-ase“ durch eine Glykosidwirkung auf der Membranaußenseite herbeigeführt werden kann.

POST hat auf einer Tagung in Prag folgende Hypothese über die Rolle der „ATP-ase“ (die er einfach als eine Membranfraktion auffaßt) vorgetragen: Der Carrier erhält durch Phosphorylierung auf der Membraninnenseite Natriumaffinität. Er transportiert dann das Natrium auf die andere Membranseite. Dort wird der Natrium-Carrier-Komplex gespalten und Natrium an die Außenlösung abgegeben. Anschließend kehrt der Carrier — nun mit Kaliumaffinität — wieder auf die Innenseite der Membran zurück. Die „ATP-ase“ wird nun dadurch aktiviert, daß ihr bei diesem Zyklus immer wieder neues Substrat angeboten wird. In einem solchen Schema kann das Herzglykosid indirekt die „ATP-ase“-Aktivität vermindern, indem es auf der Membranaußenseite die Bildung des Substrates für die „ATP-ase“ hemmt.

PASSOW: Bei den roten Blutkörperchen wird eine maximale Hemmung des aktiven Transportes erst nach $^1/_2$- bis 1stündiger Einwirkungsdauer des Strophantins erzielt (GLYNN). Wie rasch tritt der Strophantineffekt beim Nerv ein?

KEYNES (Cambridge): In a few minutes only.

PASSOW: Diese Beobachtungen könnte man am einfachsten durch die Annahme erklären, daß Strophantin beim Nerv auf der Außenseite, beim Erythrocyten dagegen auf der Innenseite der Membran wirkt. Ich glaube aber, daß unsere gegenwärtigen Kenntnisse nicht ausreichen, um diese Frage eindeutig beantworten zu können. Es dürfte aber sicher wichtig sein, die Möglichkeit in Betracht zu ziehen, daß eine Hemmung der auf der Membraninnenseite gelegenen „ATP-ase“ indirekte Folge einer Strophantineinwirkung auf der Membranaußenseite sein könnte.

HOLZER (Freiburg): Aus der Gleichheit der Michaeliskonstanten für den Transport von Natrium und Kalium darf man nicht schließen, daß es sich um dasselbe Transportsystem handelt. Ein Beispiel aus der Enzymologie ist der Umsatz von Pyruvat mit einerseits Pyruvatoxydase und andererseits Lactatdehydrogenase. In manchen Zelltypen sind die Michaeliskonstanten von Pyruvat mit diesen beiden Enzymen gleich, trotzdem sind natürlich beide Enzyme nicht identisch. — Eine zweite Frage: Kann der Austritt von Kalium aus Erythrocyten und der Eintritt von Natrium in Erythrocyten nach Zusatz von Phosphatidsäure nicht auf eine unspezifische Schädigung (Detergent-Wirkung) zurückzuführen sein, da es sich lediglich um Konzentrationsveränderungen im Sinne des Konzentrationsgefälles handelt? Dieser Versuch könnte dann nicht als Argument für die Beteiligung von Phosphatidsäuren am Kationen-Transport gewertet werden.

RUMMEL (Homburg/Saar): Es scheint nicht genügend gesichert, daß die K/Na-Pumpe so arbeitet, daß K und Na in einem Verhältnis von 1:1 ausgetauscht werden. In der Literatur sind mehrfach Werte für Na-Influx und K-Efflux zu finden, die ein zugunsten von Na verschobenes Verhältnis ergeben. Wir haben am Kaninchen-Erythrocyten den Na-Influx und K-Efflux in Abhängigkeit vom p_H gemessen. Die Resultate zeigen, daß der Na-Influx immer höher ist: bei p_H 8 wird dreimal mehr Na aus den Erythrocyten gepumpt als K hinein und bei p_H 7 rund zweimal mehr. Diese Tatsache steht in Übereinstimmung damit, daß die passive Durchlässigkeit der Membran für Natrium im entsprechenden Verhältnis höher ist als die für Kalium. Im Verteilungsgleichgewicht ist demnach der Na-Austausch rund zweimal größer als der K-Austausch.

Passow: Die mit radioaktivem Kalium oder Natrium gemessenen Ionenflüsse enthalten jeweils einen passiven und einen aktiven Anteil. Man muß daher bei der Berechnung der Relation zwischen aktiven Kalium- und Natriumverschiebungen die passiven Komponenten von den insgesamt gemessenen Ionenflüssen abziehen. Glynn hat dieser Forderung bei seinen detaillierten kinetischen Untersuchungen sehr sorgfältig Rechnung getragen, so daß mir seine Angaben besonders zuverlässig zu sein scheinen. Er findet ein Kopplungsverhältnis von 1:1. Da die Messungen aber nicht leicht exakt durchführbar sind, scheint es mir durchaus denkbar, daß verfeinerte Meßmethoden ein etwas abweichendes Verhältnis liefern werden. Bei Untersuchungen über die p_H-Abhängigkeit des Koppelungsverhältnisses müssen eventuelle Änderungen des passiven Anteils der Natrium- und Kaliumverschiebungen besonders sorgfältig berücksichtigt werden. Wenn die Anionenpermeabilität der Erythrocytenmembran tatsächlich durch dissoziable Festionen bedingt sein sollte, müßte man erwarten, daß p_H-Änderungen auch den passiven Anteil der Kationenbewegungen beeinflussen. Die Vernachlässigung dieser Änderungen müßte zu einer Verfälschung des Ergebnisses der Berechnung des Koppelungsverhältnisses führen. Wir haben ebenfalls einige Versuche über die p_H-Abhängigkeit der Kationenpermeabilität ausgeführt. Bei Erythrocyten, deren aktiver Transport durch Strophantin gehemmt worden war, beobachteten wir, daß die Geschwindigkeit der chemisch-analytisch nachweisbaren „Nettoverschiebungen" von Kalium und Natrium mit steigendem p_H zunimmt. Die von Ihnen bei Variation des p_H gefundene Änderung der Relation K^+-Efflux : Na^+-Influx könnte daher vielleicht überwiegend eine Änderung des Membranwiderstandes für die passive Diffusion der Kationen widerspiegeln und nicht eine Änderung der Relation der aktiven Komponenten von Kalium- und Natriumfluß.

Netter (Kiel): Ich möchte wegen des Kopplungsfaktors 1:1 kurz folgendes bemerken: Ich könnte mir denken, daß das Kopplungsverhältnis 1:1 betragen *muß*, damit das Zellvolumen konstant bleibt. Da die Zellmembran für Anionen ohne weiteres durchlässig ist, sollte bei anderen Kopplungsverhältnissen eine Verminderung oder Zunahme des Kationengehaltes und damit Schrumpfung oder Schwellung der Zellen erfolgen.

Passow: Tosteson und Hoffmann haben kürzlich Berechnungen publiziert, aus denen hervorgeht, daß Zellen, in denen aktiver Transport durch passive Rückdiffusion balanciert wird, ein endliches Volumen besitzen. Diesen Berechnungen wurde die Annahme zugrunde gelegt, daß aktive Kalium- und Natriumbewegungen im Verhältnis 1:1 miteinander gekoppelt sind. Die experimentelle Überprüfung der Theorie ergab recht befriedigende Übereinstimmungen zwischen Beobachtung und Berechnung. Diese Theorie sollte auch dann gelten, wenn das Kopplungsverhältnis von 1 abweicht. Man kann daher erwarten, daß sich auch unter diesen Umständen — von extremen Bedingungen abgesehen — endliche Volumina einstellen.

BADER (München): Im Hinblick auf das Modell von HOKIN möchte ich eine Bemerkung zur Markierung von Phosphatiden mit ^{32}P machen. Durchströmt man Rattenherzen mit ^{32}P, extrahiert die Phosphatide mit Chloroform-Methanol und chromatographiert sie an Kieselsäuresäulen, so bekommt man ^{32}P-Gipfel verschiedener Größe in den einzelnen Phosphatidfraktionen. Der größte ^{32}P-Gipfel ist dabei in der Fraktion, in der auch die Phosphatidsäure gefunden wird. Interessant ist, daß sich die ^{32}P-Gipfel nie genau mit den Phosphatidgipfeln decken, sondern immer etwas eher oder etwas später eluiert werden. Extrahiert man die Phosphatide von Herzen, die nicht mit ^{32}P behandelt waren, unter sonst gleichen Bedingungen und fügt man ^{32}P und Ca^{++} erst zum Chloroform-Methanol-Extrakt hinzu, so bekommt man nach Kieselsäurechromatographie beinahe die gleiche ^{32}P-Verteilung, wie wenn die Herzen perfundiert worden wären; nur der ^{32}P-Gipfel im Lecithin fehlt. Wäscht man nach Behandlung mit ^{32}P den Chloroform-Methanol-Extrakt mit Versene, so verschwindet das Ca vollständig aus der Chloroformphase und mit ihm auch der größte Teil des ^{32}P. Es sieht so aus, als ob man sich den Umweg über die Perfusion sparen könnte. Diese Ergebnisse lassen die Vermutung zu, daß anorganisches Phosphat an Phosphatid gebunden wird, wenn Ca anwesend ist. Da Phosphatide aller Voraussicht nach am Zellmembrantransport beteiligt sind, ist zu überlegen, inwieweit Resultate durch die oben beschriebenen Vorgänge bei der Extraktion der Phosphatide verfälscht werden können und ob sie vielleicht beim Transport selbst eine Rolle spielen.

HOFFMANN (Magdeburg): Ich habe eine Frage zu den ursprünglich von GARDOS beschriebenen Versuchen über die Beeinflussung des Kalium-Effluxes durch Adenosin bei Anwesenheit von Monojodessigsäure. Ist es nicht möglich, daß ein Teil der intracellulären Kaliumionen durch das freiwerdende Ammonium ersetzt wird? Ich glaube, es gibt Versuche, aus denen hervorgeht, daß Glutamin durch Erythrocyten gespalten wird, wobei es zu einer Zunahme des Kaliumefflux kommt. Dabei wird vermutlich das Kalium im Innern der Zelle durch Ammoniumionen ersetzt.

PASSOW: Die Untersuchungen über das Glutamin sind mir unbekannt. Wir haben aber bei einer Inkubation von Menschenerythrocyten mit Ammoniumionen (40 mMol/l) und einer Pentose keinen meßbaren Kaliumverlust beobachten können.

SIEBERT (Mainz): Sie haben die Frage diskutiert, ob die neueste Hokinsche Vorstellung eine stöchiometrische Beziehung zwischen Verbrauch an ATP und Transport von Natrium (und Kalium) impliziert. Kann man anhand der in den ersten beiden Vorträgen (18 bzw. 12 Na^+/Mol O_2) gemachten experimentellen Angaben die Frage entscheiden, ob stöchiometrische Bedingungen herrschen oder nicht? Mit anderen Worten: Kann man entscheiden, ob das Hokinsche System stöchiometrisch arbeitet oder nicht?

Passow: Ich glaube, daß man auf diese Frage ohne nähere Kenntnis der Eigenschaften der Zellmembran (als Lösungsmittel für die Phosphatidsäure), der Diglyceridkinase und der Phosphatidsäure keine eindeutige Antwort geben kann. Die Relation zwischen ATP-Verbrauch und aktivem Natriumtransport müßte beispielsweise p_H-abhängig sein, wenn die Phosphatidsäure in jeder ihrer beiden Dissoziationsstufen Natrium binden und durch die Membran hindurch transportieren könnte. — Ich möchte in diesem Zusammenhang nochmals sagen, daß ich kein Proponent der Hokinschen Vorstellungen sein möchte. Ich halte es für durchaus denkbar, daß Phosphatidsäurebildung und aktiver Transport zwangsläufig miteinander gekoppelt sind. Es scheinen mir aber gute Gründe dagegen zu sprechen, daß der Phosphatidsäure eine Carrier-Funktion zukommt. Es ist aber möglich, daß sie eine wichtige Rolle bei der Regulierung des Widerstandes spielt, den die Membran der passiven Rückdiffusion der Alkalimetallionen entgegensetzt.

Heinz: Wenn der von Hokin und Hokin vorgeschlagene Transportmechanismus korrekt ist, so müßte die sezernierende Membran der betreffenden Salzdrüse auf der sekretorischen Seite positiv relativ zur nutritiven Seite sein. Sollte dies inzwischen durch eine entsprechende Messung bestätigt worden sein, so würde mich interessieren, ob Ihnen irgendeine andere biologische Membran bekannt ist, bei der das elektrische Potential in dieser Weise orientiert ist. Professor Rehm (Louisville) vertrat einmal die Ansicht, daß alle bisher bekannten biologischen Membranen auf der sekretorischen Seite negativ relativ zur nutritiven seien und daß deshalb eine biologische Membran mit umgekehrter Orientierung des Potentials unwahrscheinlich sei.

Ich habe nach wie vor Schwierigkeiten, mir den Umschlag eines Carriers von der Kalium- zur Natrium-Affinität, wie er für den zwangsweisen Austausch zwischen Natrium und Kalium an verschiedenen Zellmembranen gefordert wird, vorzustellen. Wie Sie bereits hervorgehoben haben, ist auch das Hokinsche Modell der Phosphatidsäure dafür nicht geeignet. Ist Ihnen irgendein Modell bekannt, das einen derartigen Affinitätsumschlag in befriedigender Weise erklärt ?

Passow: Nein.

Dirscherl: Ich schließe die Diskussion.

Bacterial permeases

By

Adam Kepes

Service de Biochimie Cellulaire, Institut Pasteur, Paris

With 5 Figures

Introduction

Many bacteria are able to grow on a great variety of external substrates as sole carbon and energy sources. The flow of such a substrate from the medium to the interior of the cell must be considerable to satisfy the needs for the synthesis of all cell constituents and also to supply enough high energy compounds necessary for the synthetic processes. With a culture of *E. coli* on most sugar substrates, the material yield of cell mass is about 35% of the sugar consumed, thus the inflow of sugar must have a rate about three times higher than the rate of increase in cell mass. A representative value of the growth rate of an *E. coli* culture in mineral medium and a single carbon source is about 1% increase per minute. Therefore the inflow of substrate has to amount to 3% of the cell mass per minute. Although the surface area to volume ratio is extremely favourable in cells of small size, this is a very high rate of flow. With most of the good substrates this rate of uptake is realized even in very dilute solutions. The concentration of the sugar in the medium can be notably less than necessary to saturate the enzyme which performs the first metabolic transformation. If this first enzyme is in large excess above the amount which would be rate limiting for growth, a high concentration of the substrate in the medium would accelerate the reaction rate, piling up an intermediary product, e. g. glucose-1-phosphate in the case of glucose and hexokinase, and this would be a waste of glucose and ATP, decreasing the final yield. In other cases the first metabolic enzyme is not in large excess, and in this case a decrease in growth rate should be observed as the substrate concentration decreases below the level of saturation. In fact, neither a drastic decrease in material yield with high concentrations of nutrient substrate, nor a notable decrease in growth rate with low concentrations has been observed. Nor is an accumulation of intermediates dependent on the concentration of the nutrient observed unless a mutation has blocked some metabolic reaction. These observations have to be explained by some regulatory mechanism which limits the inflow of substrates at high external concentrations and accelerates it at low external concentrations.

The cells are limited by a cytoplasmic membrane which is known to be very poorly permeable for water soluble substances. Such a permeability barrier is a vital necessity in order to prevent the escape of metabolic intermediates from the cell to the medium and secure the relative constancy of the metabolic pool which reflects the adjustment of concentration of all intermediates according to the steady state reaction rates of metabolic enzymes. This relative impermeability of the cell membrane contrasts with the rate of entry of certain molecules. The narrow specificity for those molecular species which can cross the membrane cannot be explained by any purely physical model of pore size, balance of hydrophylic and hydrophobic groups in the channels, distribution of electric charges and so on. We have to postulate a mechanism having a stereospecific selectivity such as only chemical combinations are known to possess.

Definition and methodology for identification of permeases

In the course of studies on β-galactosidase induction in *E. coli*, mutant strains have been found, which could not grow on lactose as sole source of carbon. Some of these appeared to be inducible for the synthesis of β-galactosidase to the same extent as wild type strains[1]. β-galactosidase from these strains proved to be identical in every respect to normal β-galactosidase[2]. The product of its action upon lactose was a mixture of glucose and galactose; either of these two hexoses being able to support normal growth of the bacteria. The only explanation of this phenomenon was that lactose in this strain could not obtain access to its hydrolytic enzyme, β-Galactosidase, in other words that β-galactosidase was cryptic versus external substrate. Back mutations from such strains could give wild type strains lac$^+$. The barrier which prevented lactose from acceding to the β-galactosidase could not be thought of as a molecular sieve permeable to hexoses but impermeable to the larger disaccharide molecule, lactose, because the same strain could utilize maltose, another disaccharide.

The phenomenon of crypticity has been noticed in the past, and this last decade a number of new cases have been discovered.

At the same time a number of regulatory mechanisms for substrate penetration into the cells have been identified as active transport mechanisms, and called permeases[3, 4, 5].

The word permease has sometimes been criticized mainly on the basis of a misunderstanding. It has not to be taken for a mechanism which modifies cell-permeability, i. e. a physical property of the cell membrane but instead, as a mechanism which effects permeation, another word for transport.

The meaning of the word is to distinguish a specific mechanism effecting the transport of a well defined class of substrate molecules to their place of utilization. The specificity toward substrate configuration indicates that a specific cell constituent, presumably in the membrane, plays a role. It has to perform a transport without net chemical change in order to be distinct of possible exo enzymes or surface enzymes. The question is left open whether the term has to be restricted to active transport mechanisms or extended to passive transports. At the present time the permeases which have been identified *without contest* proved to perform active transport. Reciprocally, when a substrate specific active transport is demonstrated, the presence of a permease can hardly be contested, a specialized enzyme being necessary to couple an energy yielding metabolic reaction with an energy requiring transport. If active transport, namely accumulation of substrate inside the cell is not demonstrated, most of the time it is impossible to detect the unchanged substrate in the intracellular space.

In a case where the unchanged substrate is detected, inside the cell, but without accumulation, it has to be proven to be due to a specific cell constituent and not to a non-specific physical property of the membrane. To demonstrate such a specific cell constituent, the cells possessing it must be compared to cells devoid of it but otherwise identical by any criteria. Such cells can result either from a single mutation, or from an induction under appropriate conditions of growth. Another method of ascertaining the specific permeation mechanism depends on the use of substrate analogs which selectively inhibit transport.

Given a process of active transport, the experiment can fail to demonstrate an actual transport against a difference of concentration (or chemical potential) if substrate utilization is faster than penetration. In fact, this case is quite usual. For most sugars transport is the rate controlling reaction of metabolism. This difficulty can overcome in either of three ways.

1. A non utilizable analog of the substrate may be employed to demonstrate accumulation; it must competitively inhibit the uptake of the utilizable substrate.

2. A mutant cell may be found which lacks the metabolic enzyme, through which the substrate enters the metabolism. In this case the otherwise metabolizable substrate accumulates at a level above its external concentration.

3. An inhibitor may be found, which inhibits or minimizes utilization of the substrate, without inhibiting the transport process.

A fourth method, seldom used, would be to demonstrate that the apparent affinity of an intracellular enzyme for its substrate is higher in vivo than in extracts. For unequivocal demonstration at least one set of conditions has to be found where the overall reaction rate is higher in vivo than in vitro. This can only be ascertained if the extraction of the enzyme can be achieved without loss. Toluene, which destroys the permeability barrier and allows measurement of total enzyme *in situ.*, can be used, but in some instances the "decryptification" is incomplete (e. g. glucuronidase of *E. Coli*).[5]

Description of the galactoside permease

This description will serve to illustrate various methods which allow identification of the characteristics and physiological role of a permease. When a thiogalactoside is added to a suspension of either wild type induced or constitutive *E. coli*, and samples are taken at intervals to measure the intracellular concentration, the time course of uptake shown in Fig. 1 is obtained. A plateau value is reached within 2–3 min and it remains steady for at least an hour.

The experiment is commonly performed with radioactive thiomethyl galactoside (TMG), 1 ml samples being quickly filtered on a millipore filter membrane after dilution in ice-cold mineral medium, and rinsed twice with small volumes of the same medium. The radioactivity retained on the millipore is measured directly

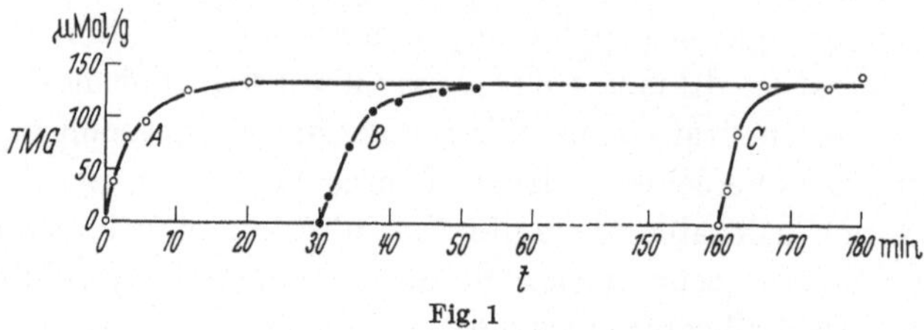

Fig. 1

after drying. The thiogalactoside content of the cells is expressed as μ moles per gram dry weight of cells. This figure must be divided by a factor of about 4 to give the concentration per ml water space in the cells. This would give for the experiment of Fig. 1 about 30 μ M per ml water space in the cells, which should be compared to the external concentration of 1 μ M/ml. The question may be raised as to what happens once the plateau is reached. Is the intracellular substrate trapped, or, is the uptake continuous with an equivalent loss resulting in a steady state? The experiment shown in Fig. 1 answers this question. Adding non radioactive substrate at zero time and the labelled substrate at various times the same rate of uptake of radioactivity is observed and the same plateau value is reached. Different thiogalactosides mutually and completely displace each other when added in sufficient excess, whereas analogous thioglucosides have no effect. The possibility of specific adsorption sites limiting the final uptake can best be excluded by the three following remarks. Although different thiogalactosides displace each other completely, the capacity of uptake is widely different for each, varying by a factor of 20 or more for the observed

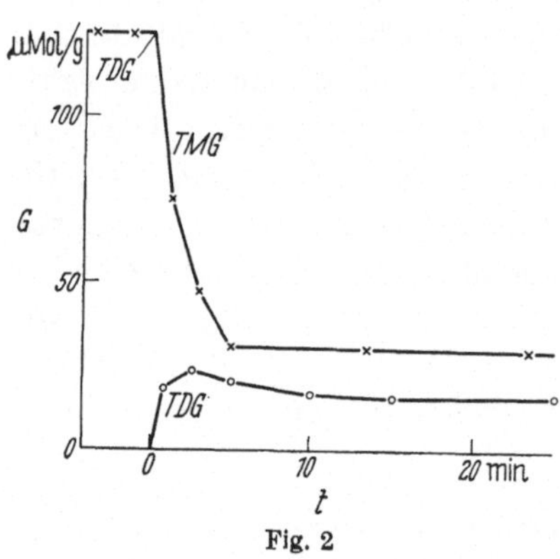

Fig. 2

extremes[7]. In a displacement experiment, when both substrates are measured simultaneously, as in Fig. 2 where TDG (thiodigalactoside) displaces TMG, more than 100 μ M/g of the second are displaced during the uptake of 15 μ M/g of the first. The third

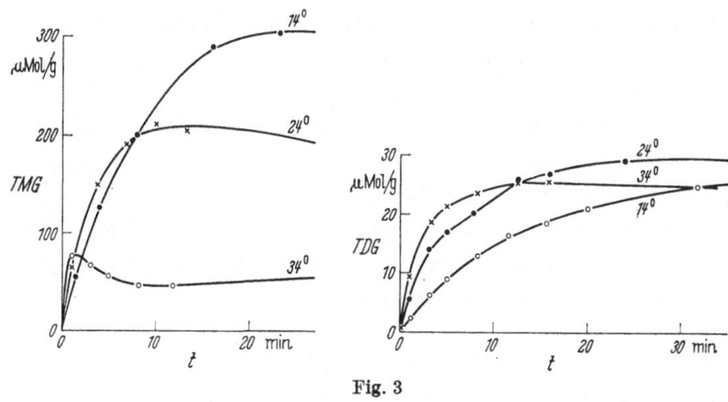

Fig. 3

argument appears in Fig. 3 where uptake of a nearly saturating concentration of TMG is measured at three different temperatures. The plateau value for the same substrate increases by a factor of 5 when the temperature is decreased from 34° C to 14° C.

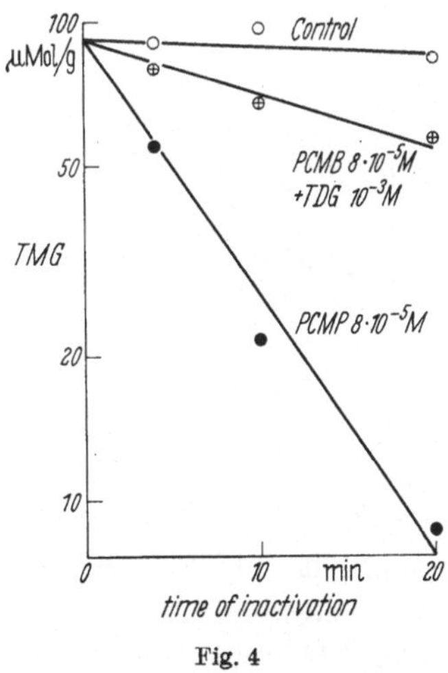

Fig. 4

Since initial rates of uptake as well, as plateau values follow Henri-Michaelis kinetics with the same apparent Km, all the described observations can be interpreted as the result of a process of active uptake balanced at steady state by an equivalent leakage, the uptake having an enzyme-like behaviour obeying Michaelis'law, the leakage having kinetics of passive transport in the range explored. If one assumes that the rate of both processes can vary independently for different substrates, different plateau values will result. If the competition takes place essentially at the site of entry, displacement of one intracellular substrate by another has no reason to follow a mole to mole ratio. Finally, if the rate constants of the entry and exit processes have a different temperature dependance, variation of the plateau value with temperature is the consequence.

The entry process is energy dependent. Oxygen consumption connected with the transport process has been measured[8] and the results are consistent with the breakdown of one mole of ATP bond per mole of TMG transported. Metabolic inhibitors like DNP or Na azide inhibit accumulation. Moreover parachloromercuribenzoate inactivates the permease and this inactivation can be slowed down by the presence of an excess of substrate as shown in Fig. 4. This strongly suggests that pCMB acts upon a structure which forms a Michaelis complex with the substrate.

Physiological significance of galactoside permease

β-galactosidase activity is routinely measured by hydrolysis of 0-nitrophenyl-β-galactoside (ONPG), a chromogenic substrate. With whole permease positive cells the hydrolysis is 15—20 times slower than with sonicated or toluenized cells, and with whole cryptic cells several hundred times slower. Consequently, for this substrate, the permease is clearly rate limiting, but by far not as drastically as simple diffusion across the membrane of cryptic cells. The objection has been raised that Na azide does not inhibit ONPG hydrolysis in permease positive whole cells although it inhibits thiogalactoside accumulation. This could be due to residual ATP synthesis. The identity of the mechanism of ONPG uptake with the accumulation of thiogalactosides is best shown by the phenomenon represented in Fig. 5 where pCMB inhibits ONPG hydrolysis in vivo to exactly the level of cryptic cell. Equilibration of cryptic cells with thiogalactosides is known not to

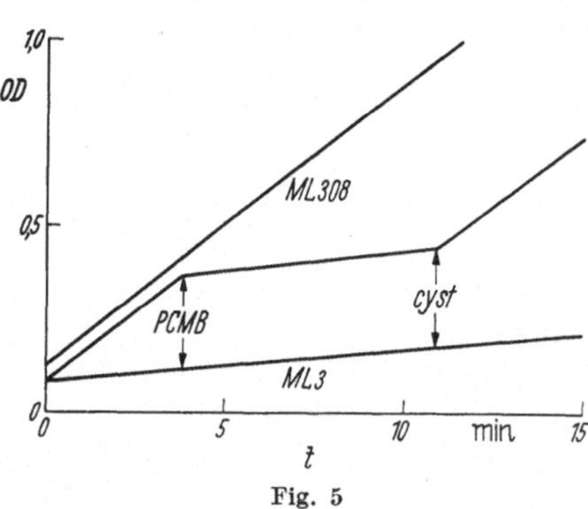

Fig. 5

be modified by pCMB. After a short period pCMB inhibition can be reversed by addition of an excess of an SH compound, cysteine in the present experiment.

Table 1 lists the known parameters for lactose accumulation by the permease positive, β-galactosidase negative, strain W 2244.

Table 1

Michaelis constant K_m	7.10^{-5}M
Plateau value at saturation*	550 μ M/g dry 137 μ M/ml cell water space
Time of half equilibrium $t^{1}/_{2}$	2.4 min
Maximal concentration ratio	$\frac{137\ 10^{-3}}{7.10^{-5}} = 1950$ fold
Maximal rate of uptake $\frac{550 \times \ln 2}{2.4\ \text{min}}$.	158 μ M/g/min 54 mg/g/min

The final calculated value of the maximal rate of uptake compares reasonably with the calculated rate of consumption of lactose when this is the sole carbon source in lac + strains: 40 mg/g/min. The higher figure is probably due to the fact that the calculations of Table 1 are based on a maximal plateau value which has been measured at a lower temperature than the normal incubation temperature.

Other bacterial permeases

In Table 2 are listed a number of permeases described in *E. Coli* together with the methods which have been used to establish their individuality. Methods numbered 1, 4, 5, 6 allow intracellular accumulation above the level in the medium. Methods of columns 2 and 3 allow only the statement of a specific mechanism of uptake without a proof that this is a transport i. e. not involving net chemical change. Even if this mechanism is a transport, methods 2 and 3 don't discriminate between the active and the passive type.

Among the various combinations of mutants helpful in demonstrating a permease arelisted only those which have lost the ability to utilize a given substrate without loss of the permease, because

* Measured at 4° C.

Table 2. *Permeases identified in Escherichia Coli. Methods employed*

Substrate	Mutants		Inducibility		Non utilizable analogs	Inhibitors of utilization	Observations	References
	Permease⁺ utilization⁻	Permease⁻ utilizing enzyme⁺ (cryptic)	Permease inducible utilizing enzyme const.	Permease constitutive utilizing enzyme inducible				
	1	2	3	4	5	6		
β-galactosides	+	+			Thiogalactosides			3
Galactose	+			+			Exit inducible	9, 10
Glucuronides		+			Thioglucuronides			5
Maltose	+							11
Glucose					α-methyl-glucoside ?			12
Ribose			+				Accumulation not demonstrated	13
Arginine, ornithine, lysine		+				Chloramphenicol		14
Phenylalanine, (tyrosine)						Chloramphenicol	Uptake generally faster than consumption	4
Methionine						Chloramphenicol		4
Valine, leucine, isoleucine						Chloramphenicol		4
Glycine, alanine, D serine		+				Chloramphenicol		14, 15
Histidine		+				Chloramphenicol		15
Proline		+				Chloramphenicol		15, 16
K					Not metabolized			17

they perform accumulation, and those which have lost the permeation mechanism without loosing the metabolic enzymes of utilization, because they prove that permease character is distinct from metabolic enzymes.

Inducibility also has been listed only in those cases when it is different for permease and for the enzymes involved in utilization, for the same reasons as above.

Galactose permease is distinct from galactoside permease genetically and functionally. Galactose permease is synthesized constitutively by gal^- strains. Both rate of uptake and plateau value may be described by Michaelis equations with the same Km. One peculiarity is the increase of the rate of exit when bacteria are grown on low concentrations of galactose. This seems to indicate that the exit process is not simple diffusion, although is behaves as a passive process.

Glucuronide permease is inducible together with the metabolic enzyme glucuronidase. Cryptic mutants have been found. The initial rate of uptake obeys Michaelis' law. The plateau values show some deviation which could be interpreted as a rate of exit more than proportional to the concentration difference and could be due to the electrostatic charge of the substrate.

Maltose permease is inducible in mutants which do not utilize maltose. Although some of the maltose taken up is incorporated in insoluble cell constituents, the major part is exchangeable, and can be recovered as unchanged maltose. Published data do not state Michaelis type concentration dependence.

Glucose permease is rather difficult to detect. Some mutants of *E. coli* unable to use glucose have been isolated and reversible accumulation of glucose have been shown in some instances, but rapid reversion to a $glucose^+$ condition made these experiments non reproducible. On the other hand, all strains of *E. coli* accumulate α-methyl-glucoside constitutively. As α-glucosides are not metabolized, the physiological significance of such a system is open to question. It may be that the normal role of the α-glucoside permease is to take up glucose, a constitutive characteristic of all strains.

Ribose uptake is inducible in *E. coli* strains whereas ribose is constitutively metabolized, when released intracellularly by hydrolysis of purine ribosides. This fact strongly suggests the existence of a ribose permease.

All aminoacid permeases are constitutive in *E. coli*. They are evidenced when incorporation is inhibited by chloromycetin. Some complication may arise for these aminoacids which are actively metabolized in addition to simple incorporation into proteins. Nevertheless at least seven different permeases have been individually characterized, each for a group of structurally related aminoacids. The group valine leucine isoleucine illustrates the fact that the specificity of a permease is independant of the biosynthetic "families". Although it is difficult to select for aminoacid-permease negative mutants, this has been achieved for four different systems either by resistance to amino acid analogs or by a more complicated penicillin technique.

A potassium permease has been recently described in *E. coli*. Negative mutants cannot grow on low K^+ media. The genetic locus of this mutation has been mapped. Permease positive strains accumulate K^+ up to a hundred-fold concentration ratio versus medium. The kinetics of the accumulation obey Michaelis' law. The mechanism is completely unspecific toward the anion which must be transported to respect electrical neutrality. K^+ is very probably present mostly as the free ion in the protoplasm and cannot be metabolized in the usual sense. Alkali metal ion transport has a tremendous importance in any living cell, and has been for a long time the main subject in studies of active transport. The finding in *E. coli* of a system governed by a single gene and thus interpreted as the function of a single specific protein component puts this ubiquitous "vital function" on the same level as other transport systems, specific for organic molecules.

This fact raises the question of the mechanism of the transport processes. The generally accepted theory is that active transport depends on diffusion of an intermediate, generated on one side of the membrane and returned to the original molecular species on the other side. At least one of these two reactions is mediated by an enzyme. If two enzymes are present, either one could perform the coupling with an energy yielding metabolic reaction.

This hypothetical intermadiate is usually visualized as a complex with a carrier molecule. The concept of carrier, however, is rather vague. It might be a highly specific molecule or a ubiquitous small radical, it even could be an absence of a radical if the intermediate resulted from dehydration of the transported molecule. The main question is whether the combination with the hypothetic carrier involves covalent linkage within the transported molecule. This seemed highly probable with organic molecules, but this has not yet been demonstrated. On the other hand secondary valencies, electrostatic, hydrophobic forces and hydrogen bonding could account for the specificity and for the relative stability of the intermediate. Hydrogen bonding has been advocated by STEIN and DANIELLI in the case of the glycerol transporting system of red blood cells, electrostatic forces must play a major role in the case of alkali cations; and still in both of these cases a high specificity is observed. Thus the intervention of such forces in permease systems for organic molecules must be contemplated as an alternative to covalent forces. In this case it becomes difficult to visualize the permease as an enzyme. Whatever be the nature of the forces in the intermediate, a Michaelis type interaction of the transported substrate with a specific enzyme constituent must be postulated. In this Michaelis complex the transported substrate serves either as a substrate for the enzyme if it participates in covalent bond changes, or as an enzyme activator, if the enzyme performs *covalent work* on the energy donor molecule only. This would explain the triggering of oxygen consumption by addition of TMG to a galactoside permease positive cell suspension.

References

[1] DEERE, C. J., A. D. DULANEY and I. D. MICHELSON: J. Bact. **37**, 355 (1939).
[2] MONOD, J., and M. COHN: Advanc. Enzymol. **13**, 67 (1952).
[3] COHEN, G. N., et H. V. RICKENBERG: Ann. Inst. Pasteur **97**, 693 (1956).
[4] RICKENBERG, H. V., G. N. COHEN, G. BUTTIN et J. MONOD: Ann. Inst. Pasteur **91**, 829 (1956).
[5] STOEBER, F.: C. R. Acad. Sci. (Paris) **244**, 1091 (1957).
[6] KEPES, A., et J. MONOD: C. R. Acad. Sci. (Paris) **244**, 809 (1957).
[7] KEPES, A.: Biochim. biophys. Acta **40**, 70 (1960).
[8] KEPES, A.: C. R. Acad. Sci. **244**, 1550 (1957). (Paris)
[9] HORECKER, B. L., J. THOMAS and J. MONOD: J. biol. Chem. **235**, 1580 (1960).
[10] HORECKER, B. L., J. THOMAS and J. MONOD: J. biol. Chem. **235**, 1587 (1960).
[11] WIESMEYER, H., and M. COHN: Biochim. biophys. Acta **39**, 440 (1960).
[12] COHEN, G. N., and J. MONOD: Bact. Rev. **21**, 169 (1957).
[13] EGGLESTON, L. V., and H. A. KREBS: Biochem. J. **73**, 264 (1959).
[14] SCHWARTZ, J. H., W. K. MAAS and E. J. SIMON: Biochim. biophys. Acta **32**, 582 (1959).
[15] LUBIN, M., D. H. KESSEL, A. BUDREAU and J. D. GROSS: Biochim. biophys. Acta **42**, 535 (1960).
[16] BRITTEN, R. J., R. B. ROBERTS and E. F. FRENCH: Proc. nat. Acad. Sci. (Wash.) **41**, 863 (1955).
[17] LUBIN, M., and D. KESSEL: Biophys. biochim. Res. Com. **3**, 67 (1960).

Zuckertransporte

Von

Walter Wilbrandt

Pharmakologisches Institut der Universität Bern

Mit 14 Textabbildungen

Die Transportsysteme, mit denen Zucker durch Zellmembranen und durch lebende Zellen hindurchtransportiert werden, zeichnen sich vor denjenigen für Aminosäuren und Ionen durch einige Eigentümlichkeiten aus.

Erstens sind an mehreren Zellarten, die günstige Bedingungen für die experimentelle Analyse bieten, die Systeme nicht befähigt, bergauf, d. h. entgegen dem Konzentrationsgradienten, zu transportieren. Trotzdem bestehen überzeugende Hinweise dafür, daß auch in diesen Fällen die Zuckerverschiebungen nicht durch einfache passive Diffusion, etwa durch poröse Membranen, erfolgen, und vieles spricht dafür, daß Trägermechanismen beteiligt sind[29, 30, 45, 53, 56, 61]. Die Beobachtungen an solchen Zellen bieten daher eine willkommene Gelegenheit, die Eigentümlichkeiten des Trägerprinzips per se zu studieren, unabhängig von den Voraussetzungen für Aufwärtstransporte, d. h. von den Einrichtungen der „Pumpen“. Diese Analyse hat zu einer Reihe von Charakteristika geführt, die sich quantitativ aus dem Trägerprinzip ableiten lassen und in guter Übereinstimmung mit experimentellen Beobachtungen stehen.

Einige Beobachtungen dieser Art sind auch an Systemen gemacht worden, die bergauf zu transportieren vermögen. Das legt die Vermutung nahe, daß auch in diesen Fällen ähnliche Prinzipien benützt werden, und führt zu der Frage, welche zusätzlichen Elemente geeignet sein können, aus einem Trägersystem, das zunächst nur zu Konzentrationsausgleich führt (Ausgleichssystem), eine „Pumpe“ zu machen.

Eine Eigenheit der Zucker als Transportsubstrate, die die Analyse erleichtert, ist das Fehlen von freien elektrischen Ladungen. Im Gegensatz zu den Transportsystemen für Aminosäuren und

Ionen muß daher das Membranpotential nicht als zusätzliche treibende Kraft und damit als komplizierender Faktor in Betracht gezogen werden.

Nachteilig ist dagegen, daß in vielen Fällen die Geschwindigkeit der Stoffwechselvorgänge, denen die eintretenden Zuckermoleküle unterworfen werden, im Verhältnis zur Penetrationsgeschwindigkeit so groß ist, daß ein annähernder Konzentrationsausgleich nicht erreicht wird, was die Analyse erschwert. Eine Maßnahme, mit der dieser Schwierigkeit (ähnlich wie im Falle der Aminosäuren) mit Erfolg begegnet werden konnte, ist die Verwendung solcher Zuckerarten oder Zuckerderivate, die vom Stoffwechsel nicht erfaßt werden.

Ein Punkt, der den Zuckertransportsystemen besonderes Interesse verleiht, ist die Tatsache, daß an einigen von ihnen Wirkungen des Insulins nun von einer größeren Anzahl von Autoren nachgewiesen worden sind. Es liegen bereits Versuche vor, die Insulinwirkung in quantitativer Weise mit Veränderungen der Parameter der Transportsysteme in Beziehung zu setzen. Offenbar kann mindestens ein Teil der biologischen Insulinwirkungen mit diesen Effekten in Zusammenhang gebracht werden.

1. Die wichtigsten bisher studierten Systeme

Die bisher studierten Transportsysteme für Zucker sind teils solche, die Substratverschiebungen zwischen Zellinnerem und Umgebung durch die Zellmembran hindurch bewirken (celluläre Systeme) z. T. solche, die durch Epithelzellschichten hindurch transportieren (transcelluläre Systeme).

Von den *cellulären Systemen* bewerkstelligen die meisten lediglich Ausgleichstransporte. Die am extensivsten untersuchte Zellart ist der Erythrocyt. Im Zusammenhang mit dem Insulinproblem sind vor allem quergestreifte Muskeln und Herzmuskeln untersucht worden[22, 37, 38, 39, 40, 41]. In neueren Studien wurden außerdem verschiedene Tumorzellen[10, 35, 36], die Zellen der Leber[3], Lymphocyten[23], schließlich Hefezellen[2,6] und Bakterien[7, 25, 26, 27] herangezogen. Unter all diesen Zellarten ist bisher nur in einem Fall, nämlich bei den Permease-Systemen von Bakterien, über die Kepes im vorhergehenden Vortrag ausführlich berichtet hat, von Bergauf-Transporten berichtet worden. Die übrigen cellulären Systeme sind Ausgleichssysteme.

Dagegen sind die *transcellulären Transporte* durch Nierenepithelzellen und Darmepithelzellen[1] potentiell „Pumpen", d. h. in beiden Fällen kann der Zucker bergauf, entgegen der Konzentrationsdifferenz, transportiert werden.

2. Hinweise auf Trägermechanismen und Charakteristik von Trägertransporten in Ausgleichssystemen

Daß die Zuckerverschiebungen durch Zellmembranen und durch resorbierende Zellen nicht durch einfache passive Diffusion bedingt sein können, wurde frühzeitig wahrscheinlich aus Beobachtungen über die hohe strukturelle Spezifität dieser Systeme und aus Beobachtungen über die Transportkinetik.

a) Spezifität. Beobachtungen über Strukturspezifitäten des Zuckerdurchtritts durch Zellmembranen, die mit einfacher Diffusion nicht in Einklang zu bringen sind, wurden schon von Kozawa 1914 am Erythrocyten mitgeteilt und führten zu der bemerkenswerten Vermutung Höbers, daß spezifische Transporteinrichtungen bestehen, die möglicherweise dem Einfluß von Insulin unterliegen. Während am Erythrocyten sich später die Annahme von Insulinwirkungen nicht bestätigt hat, steht eine solche Wirkung an anderen Zellen heute in der Diskussion des Wirkungsmechanismus des Insulins im Vordergrund.

Von späteren Beobachtungen auffälliger Strukturspezifitäten seien erwähnt das unterschiedliche Verhalten von Aldosen und Ketosen bei Erythrocyten[52] und die großen Unterschiede in der Penetrationsgeschwindigkeit stereoisomerer Aldosen, insbesondere der optischen Antipoden gleicher Zucker[55].

Beispiele besonders eindrucksvoller struktureller Spezifität ergaben Beobachtungen über die sog. „kryptischen" Stämme von Bakterien, die bestimmte Substrate nicht abzubauen vermögen, obwohl in den Zellen alle dafür notwendigen Enzyme vorhanden sind (vgl. die vorhergehende Darstellung von Kepes). Doudoroff[14] hat z. B. einen Bakterienstamm beschrieben, der Glucose nicht vergärt, wohl dagegen Maltose. Da der Abbau der Maltose mit der hydrolytischen Spaltung des Moleküls in zwei Glucosemoleküle beginnt, ist die Zelle in diesem Fall offenbar imstande, die so entstandenen Glucosemoleküle abzubauen, nicht dagegen von außen zugesetzte. Die daraus sich ergebende Schlußfolgerung, daß die

Zellmembran für das große Maltosemolekül durchlässig, für das kleinere Glucosemolekül dagegen undurchlässig ist, ist mit einer Passage durch poröse Membranen unvereinbar und zwingt zur Annahme spezifischer Transportsysteme.

Lefèvre und Marshall haben als Grundlage der Spezifität bei den Aldosen an Menschenerythrocyten angenommen, daß die sog. Cl-Konformation des Pyranoserings (Abb. 1) die für den Transport günstige Konfiguration darstellt. Sie konnten für 14 Aldosen zeigen, daß die Abstufung der nach Reeves berechneten Stabilitätsbedingungen für die Cl-Konformation mit derjenigen der Penetrationsgeschwindigkeiten parallel geht, und deuteten diese Beziehung mit entsprechenden Abstufungen der Affinität der Zuckermoleküle zu einem Trägersystem.

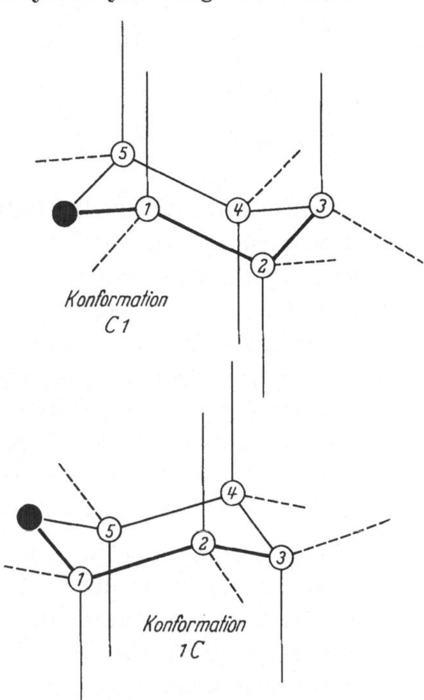

Abb. 1. Die beiden von Lefèvre und Marshall diskutierten Sesselkonformationen C1 und C1, von denen nach den Autoren die Konformation C1 die für den Transport günstige Form darstellt. Den verschiedenen Wannenformen, die noch denkbar sind, wird zu wenig Stabilität zugeschrieben, als daß sie in Betracht gezogen werden müßten. (Nach Lefèvre und Marshall 1958)

Für die nähere Charakterisierung möglicher Reaktionen zwischen Zuckersubstrat und supponiertem Träger ist von besonderem Interesse die Entscheidung, ob bestimmte Hydroxylgruppen des Zuckermoleküls für den Transport unentbehrlich sind. Das extensive Studium verschiedenartiger Zuckerderivate, insbesondere methylierter Zucker und verschiedener Deoxyzucker, hat in jüngerer Zeit zu Resultaten geführt, die eine partielle Beantwortung dieser Frage für mehrere Zellarten gestatten. Rosenberg[44] hat vor kurzem die Resultate zusammengestellt und diskutiert. Seine Darstellung ist in Tab. 1 wiedergegeben. Sie zeigt, daß nur bei einem der studierten

Tabelle 1. *Entbehrlichkeit der einzelnen Hydroxylgruppen von Glucose für 4 Transportsysteme*

C-Atom	Menschen-Erythrocyten	Penetration in		Akkumulation im Hamster-Darm
		Ascites-Tumor-Zellen	Muskel-Zellen	
C1	E	E	?	E
C2	E	E	E	U
C3	E	E	E	E
C4	E	E	E	E
C6	E	E	E	E

(Nach ROSENBERG 1960). E = entbehrlich, U = unentbehrlich.

Systeme bisher eine individuelle Hydroxylgruppe als unentbehrlich erkannt worden ist: Die Hydroxylgruppe am Kohlenstoff 2 im Falle des Bergauftransportes durch die Darmepithelzellen. In allen anderen untersuchten Fällen hat sich die Entfernung oder die Blockierung einzelner Hydroxylgruppen zwar quantitativ auf die Transportgeschwindigkeiten ausgewirkt, dagegen nicht qualitativ den Transport verunmöglicht. ROSENBERG schließt daraus, daß an der Reaktion zwischen Zucker- und Trägermolekül mehrere Hydroxylgruppen beteiligt sind und schlägt eine Verknüpfung durch eine Mehrzahl von Wasserstoffbindungen vor.

b) Kinetik. Frühe kinetische Befunde, die darauf hinwiesen, daß der Zuckerdurchtritt durch Zellmembranen nicht auf freier Diffusion beruhen kann, waren die Beobachtung von EGE[16], daß der Eintritt von Zucker in Menschenerythrocyten aus niedrigen Konzentrationen außerordentlich rasch, aus hohen Konzentrationen dagegen sehr langsam erfolgt. Ferner wurde an den gleichen Zellen mehrfach gefunden[34,60], daß die Werte der Permeabilitätskonstanten für Glucose, die auf Basis der Diffusionskinetik aus experimentellen Beobachtungen errechnet werden, bei verschiedenen Substratkonzentrationen keineswegs konstant sind, sondern um mehrere Größenordnungen variieren können, was die Unrichtigkeit der für die Berechnung verwendeten Voraussetzungen erweist.

Dagegen stehen quantitative Schlußfolgerungen, die sich kinetisch aus der Annahme eines Trägermechanismus ergeben, in mehrfacher Hinsicht in guter Übereinstimmung mit experimentellen Beobachtungen.

Abb. 2 zeigt schematisch den *Mechanismus*. Das Substrat S reagiert auf der ersten Membranseite mit dem Träger C und der gebildete Komplex CS passiert die Membran. Die Membranpassage kann nach der einfachsten möglichen Annahme durch Diffusion erfolgen, es gibt aber eine Reihe anderer Möglichkeiten, wie Molekülrotation oder dgl., die kinetisch gleiche Konsequenzen haben. CS spaltet sich auf der andern Seite unter Freisetzung von S, das in die wäßrige Phase übertritt, und von C, das in der Membran bleibt und durch Diffusion zur Seite 1 zurückkehrt.

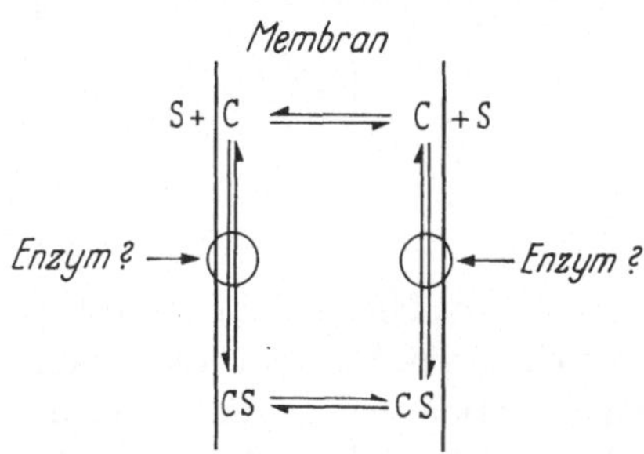

Abb. 2. Trägerschema für den ausgleichenden Zuckertransport durch Zellmembranen. Das Substrat S reagiert reversibel mit dem Träger C, passiert die Membran in Form des Trägersubstratkomplexes CS und wird auf der anderen Seite, wiederum in reversibler Reaktion, freigesetzt, worauf der Träger C zurückdiffundiert

Für die Reaktion zwischen Substrat und Träger ist, vor allem in Hinblick auf die hemmende Wirkung zahlreicher Enzyminhibitoren, enzymatische Katalyse angenommen worden[45]. Kinetisch liegen dafür bisher allerdings nur wenige Hinweise vor [63]. Möglicherweise ist die Enzymbeteiligung nicht obligatorisch. Besonders wahrscheinlich ist sie jedoch im Fall der Permeasen, auf den zurückzukommen sein wird.

Das in Abb. 2 gezeigte System bewirkt Nettotransport nur bis zum Konzentrationsausgleich. Solange $S_1 \neq S_2$, besteht in der Membran ein Gradient für CS und ein Gegengradient für C. Beide Gradienten verschwinden, wenn $S_1 = S_2$, wie in der später zu besprechenden Abb. 6 erkennbar ist.

Die *Transportkinetik* für ein solches System ist unter verschiedenen Voraussetzungen berechnet worden. Die einfachsten Beziehungen ergeben sich, wenn für die Reaktion zwischen S und C Gleichgewicht angenommen wird. In diesem Fall führt die Beteiligung von Enzymen in der in Abb. 2 angegebenen Weise zu keinen kinetischen Konsequenzen. Dagegen verändert sie die Kinetik dann, wenn die Reaktionsgeschwindigkeiten limitieren, d. h. die Reaktion nicht im Gleichgewicht ist. Der wesentliche Unterschied besteht darin, daß unter diesen Bedingungen die unten zu besprechende

E-Kinetik nur bei enzymatischer Katalyse zu erwarten ist. Da sie am Erythrocyten mehrfach gefunden wurde, ist in diesem Fall entweder Gleichgewicht der Reaktion oder enzymatische Katalyse anzunehmen.

Im einfachen Fall des Reaktionsgleichgewichtes ergibt sich für die Transportgeschwindigkeit[46,50] die folgende Beziehung zu den Substratkonzentrationen S_1 und S_2:

$$V = D' C_t \left(\frac{S_1}{S_1 + K_m} - \frac{S_2}{S_2 + K_m} \right) \tag{1}$$

oder

$$V = V_{max} K_m \frac{(S_1 - S_2)}{(S_1 + K_m)(S_2 + K_m)}, \tag{1a}$$

in der K_m die Michaelis-Konstante des Träger-Substrat-Komplexes CS, V_{max} die Maximalgeschwindigkeit, C_t die gesamte Trägerkonzentration und D' die Permeabilitätskonstante des Komplexes bedeutet. Die beiden Terme in Gl. (1) sind als Kapazitäts- und Affinitätsterm bezeichnet worden (V_{max} als Kapazitäts- und K_m als Affinitätsparameter). Extensive experimentelle Prüfungen dieser Gleichung liegen bisher vor für drei Sonderfälle, die sich unter vereinfachten Annahmen ergeben.

Der erste Sonderfall tritt dann ein, wenn in der Klammer der Gl. (1) das zweite Glied gegenüber dem ersten vernachlässigt werden kann, mit andern Worten, wenn der Sättigungsgrad des Trägers auf der Seite 1 der Membran einen endlichen, auf der Seite 2 dagegen einen zu vernachlässigenden Wert besitzt. In diesem (und nur in diesem) Fall, ergibt sich quantitative Übereinstimmung mit der Michaelis-Mentenschen Gleichung, wie sie mehrfach beschrieben wurde. Die dafür notwendigen Voraussetzungen können beispielsweise geschaffen werden durch rasche Entfernung des Substrats auf der Membran-Seite 2 (vor allem durch Stoffwechselvorgänge) oder durch Veränderung von K_m auf der Seite 2 im Zusammenhang mit energieliefernden Stoffwechselreaktionen, worauf unten zurückzukommen sein wird. Als Beispiel zeigt Abb. 3 den Eintritt von Glucose in den perfundierten Rattenherzmuskel[39]. Weitere Beispiele sind entsprechende Beobachtungen am Darmepithel[9,19], an Tumorzellen[10] und an Bakterien[7,27].

Zwei weitere Sonderfälle hängen vom numerischen Verhältnis zwischen den Substratkonzentrationen einerseits und der Michaelis-Konstante K_m andererseits ab. Sind die Substratkonzentrationen klein im Verhältnis zu K_m, so ist die Kinetik, wie aus Gl. (1a) hervorgeht, von derjenigen der freien Diffusion nicht zu unterscheiden:

$$V = V_{max} \frac{I}{K_m} (S_1 - S_2) . \tag{2}$$

Sie wurde daher als D-Kinetik bezeichnet. Daraus ergibt sich die (nicht immer beachtete) Konsequenz, daß Diffusionskinetik einen Trägermechanismus nicht ausschließt. Die Transportgeschwindigkeit ist der Dissoziationskonstante K_m umgekehrt proportional, m. a. W., sie ändert sich symbat mit der Affinität.

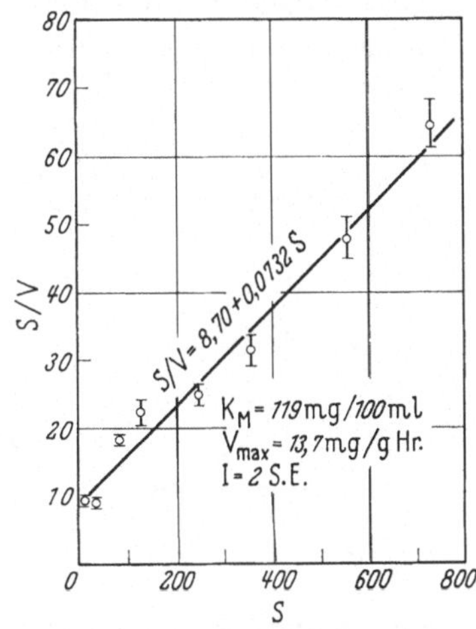

Abb. 3. Versuche über den Eintritt von Glucose in den perfundierten Herzmuskel, dargestellt nach LINEWEAVER und BURK. S = Zuckerkonzentration, V = Penetrationsgeschwindigkeit. (POST, MORGAN und PARK 1961)

Bis hierher ähneln die kinetischen Konsequenzen des Trägertransports durchaus denjenigen einer enzymatischen, der Michaelis-Mentenschen Gleichung folgenden Reaktion. Der Parallelismus schwindet aber weitgehend, und neue, auf den ersten Blick unerwartete Beziehungen ergeben sich für die Transportkinetik unter der zweiten Grenzfallbedingung, nämlich hoher Substratkonzentrationen im Verhältnis zu K_m, d. h. hochgradiger Sättigung des Trägers auf beiden Seiten der Membran.

Die erste Konsequenz aus dieser Relation betrifft die Abhängigkeit der Transportgeschwindigkeit von den Substratkonzentrationen. Sie wird, wie sich ebenfalls aus Gl. (1a) für diese Bedingung leicht ableiten läßt, proportional der Differenz der beiden reziproken Substratkonzentrationen:

$$V = V_{max} K_m \left(\frac{1}{S_2} - \frac{1}{S_1}\right) . \tag{3}$$

Diese Kinetik wurde als E-Kinetik bezeichnet. Sie ist eine Sättigungskinetik, deren Besonderheit darin liegt, daß die Maximalgeschwindigkeit (bei hohem S_1) in ausgesprochener Weise von S_2 abhängt. Besonders frappant ist diese Abhängigkeit im Gebiet niedriger Werte von S_2, in dem die Größe $S_1 - S_2$ (die im Falle freier Diffusion geschwindigkeitsbestimmend wäre) sich praktisch nicht ändert. Abb. 4 zeigt entsprechende Beobachtungen am Erythrocyten.

Für die Regulation der Eintrittsgeschwindigkeit von metabolisierbaren Zuckern (vor allem Glucose) in Zellen mit lebhaftem Stoffwechsel (der S_2 niedrig hält) bietet diese Kinetikform offensichtlich besondere Vorteile.

Experimentelle Resultate an Erythrocyten ergaben für Glucose in zwei unabhängigen Untersuchungsserien vorzügliche Übereinstimmung mit der E-Kinetik[51, 59]. Auch die Forderung, daß bei gleicher Konzentration zweier verschieden affiner Zucker der höher affine mit E-Kinetik, der schwächer affine mit D-Kinetik transportiert wird, ist für Glucose und Sorbose[51] sowie für Glucose und Fructose[53] erfüllt. Daß die Penetrationskinetik der Glucose, die bei hohen Konzentrationen dem E-Typ folgt, bei niedrigen Konzentrationen in den Typ D übergeht (wie zu fordern), wird daraus wahrscheinlich, daß die numerischen Werte experimentell ermittelter „Permeabilitätskonstanten" sich in diesem Konzentrationsgebiet tatsächlich der Konstanz nähern, wie Abb. 5 zeigt.

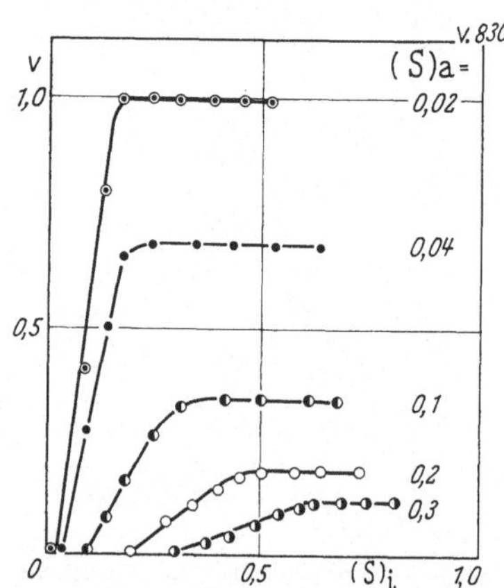

Abb. 4. Charakteristik der E-Kinetik nach Austrittsversuchen mit Glucose an Menschenerythrocyten. v = Austrittsgeschwindigkeit, $(S)_a$ = Außenkonzentration, $(S)_i$ = Innenkonzentration. Versuche bei 37° C. (WILBRANDT 1957)

Die zweite Konsequenz hoher Sättigung betrifft die Beziehung zwischen *Transportgeschwindigkeit und Affinität.* Wie Gl. (3) für hohe Sättigung zeigt, wird unter dieser Bedingung die Transportgeschwindigkeit nicht dem reziproken Wert der Konstante K_m proportional, sondern der Konstante K_m selbst. Die Geschwindigkeit nimmt dann also mit steigender Affinität nicht zu, sondern ab. Es läßt sich zeigen, daß der Übergang zwischen der ansteigenden und der abfallenden Abhängigkeit gegeben ist durch die Beziehung

$$K_m^2 = S_1 S_2 . \qquad (4)$$

Aus diesen Beziehungen ergibt sich die Konsequenz, daß Substrate mit verschiedener Affinität zu einem Trägersystem sich in der Transportgeschwindigkeit bei niedrigen Konzentrationen umgekehrt abstufen müssen als bei hohen. Diese Voraussage hat sich bestätigt in Versuchen an Erythrocyten[54], an denen die Geschwindigkeitsfolge von 5 Zuckern zwischen den Konzentrationen 0,03 M und 1,5 M umkehrte (Abb. 5).

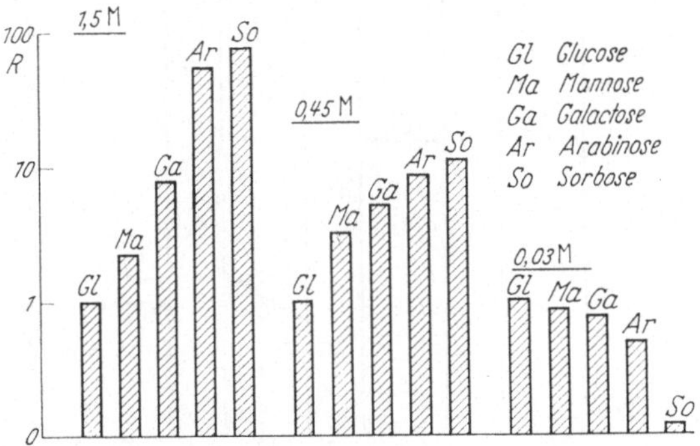

Abb. 5. Penetrationsgeschwindigkeit von 5 Zuckerarten an Menschenerythrocyten, bei 37° C in Abhängigkeit von der Zuckerkonzentration. R = relative Penetrationsgeschwindigkeit (Glucose = 1 gesetzt) (WILBRANDT 1956)

c) Der Gegentransport[43, 47, 50, 56, 64]. Der angenommene Trägermechanismus hat zwei wesentliche Elemente: die Bindung des Transportsubstrats an einen Komplexpartner und die Bewegung des gebildeten Komplexes durch die Membran (beispielsweise durch Diffusion). Erst das zweite Element, die Mitführung des Substrates durch den Träger bei der Bewegung zur jenseitigen Membranfläche, rechtfertigt die Bezeichnung „Trägermechanismus" im eigentlichen Sinne. Die bisher abgeleiteten Beziehungen stehen jedoch alle nur in Beziehung zum ersten Element, zur Bindung des Substrates an eine Membrankomponente. Sie sind, unter geeigneten Bedingungen, in quantitativ gleicher Weise ableitbar für Systeme, bei denen der Komplexpartner ein fixer Bestandteil der Membran ohne Bewegungsfreiheit ist und bei denen das Substrat sich in Sprüngen von einer Bindungsstelle zur andern bewegt.

Für die Bewegung des Komplexes durch die Membran gibt es aber ein Kriterium, das von solchen Systemen nicht geteilt wird: den Gegentransport. Darunter wird ein Transport verstanden, der bergauf erfolgen kann und der nicht getrieben wird von der Konzentrationsdifferenz des transportierten Substrats, sondern von der gleichzeitigen Bewegung eines zweiten Substrats für den gleichen Träger in der Gegenrichtung.

Gegentransport. Gradientenschema

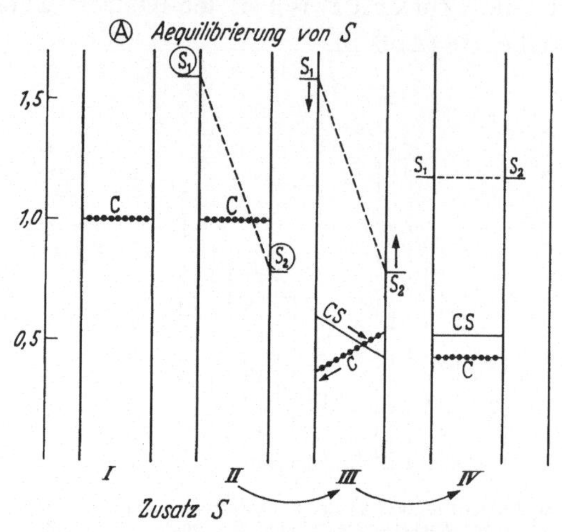

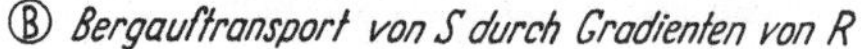

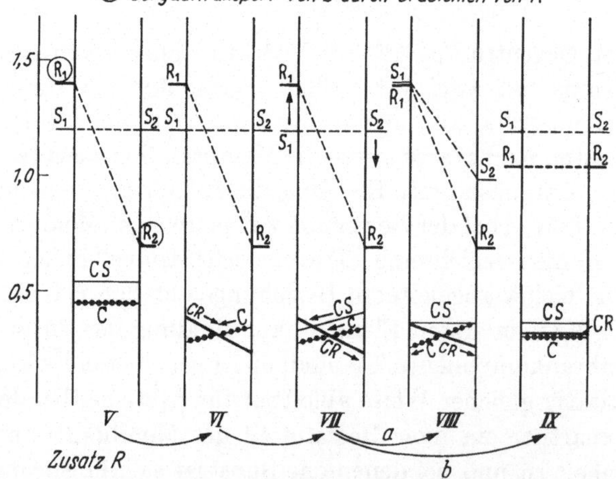

Abb. 6. Gradientenschema für das Zustandekommen des Gegentransports in einem ausgleichenden Trägersystem. Näheres im Text

Abb. 6 zeigt das Zustandekommen eines solchen Gegentransports in anschaulicher Form durch die entstehenden Gradienten.

Der erste Teil der Abbildung zeigt den Ausgleichstransport eines ersten Substrats S, wie er oben beschrieben wurde. Im zweiten Teil induziert dann der Zusatz eines zweiten Substrats R ($R_1 > R_2$) durch ungleiche Reaktion mit C auf den beiden Membranseiten einen Gradienten für CR und, als sekundäre Folge, einen Gegengradienten von C. Dieser Gegengradient induziert, obwohl $S_1 = S_2$, einen Gradienten für CS, der zu einem Bergauftransport von S führt.

Das weitere Geschehen hängt dann davon ab, ob die Konzentrationsdifferenz von R aufrecht erhalten wird oder ob sie sich ausgleicht. Im ersteren Fall wird S bergauf transportiert, bis ein neuer stationärer Zustand erreicht wird, bei dem

$$\frac{S_1}{S_2} = \frac{R_1' + 1}{R_2' + 1}\,. \tag{5}$$

Diese Entwicklung ist in Abb. 6 unter VIII wiedergegeben.

Gleicht die Konzentrationsdifferenz von R sich aus, so ist der Gegentransport temporär und nach einiger Zeit sind alle Gradienten in der Membran verschwunden (IX in Abb. 6).

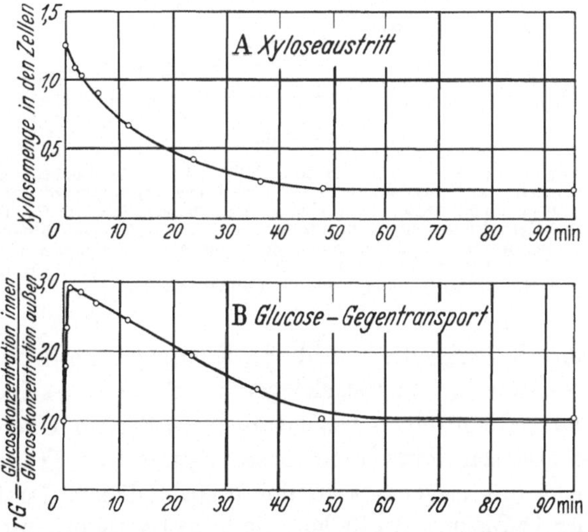

Abb. 7. Beispiel eines Gegentransports von Glucose an Menschen-Erythrocyten. Die Zellen wurden zunächst äquilibriert mit einer Lösung, die Xylose und Glucose enthielt und dann in eine Lösung mit gleicher Glucosekonzentration, aber niedrigerer Xylosekonzentration überführt, so daß Xylose aus den Zellen austritt. Gleich zu Beginn des Xyloseaustritts wird Glucose entgegen ihrem Gradienten in die Zellen transportiert (Gegentransport) und die ungleiche Verteilung der Glucose hält an, bis der Xyloseaustritt abgeschlossen ist

Der Gegentransport ist von WIDDAS[50] abgeleitet und für Zucker mehrfach am Erythrocyten[37, 47], ferner am Herzmuskel[38] und an Hefezellen[2, 6] demonstriert worden. Abb. 7 zeigt eine Beobachtung an Erythrocyten. Für diese Zellarten ist also neben der Bindung des Substrats eine Bewegung des gebildeten Komplexes durch die Membran hindurch anzunehmen.

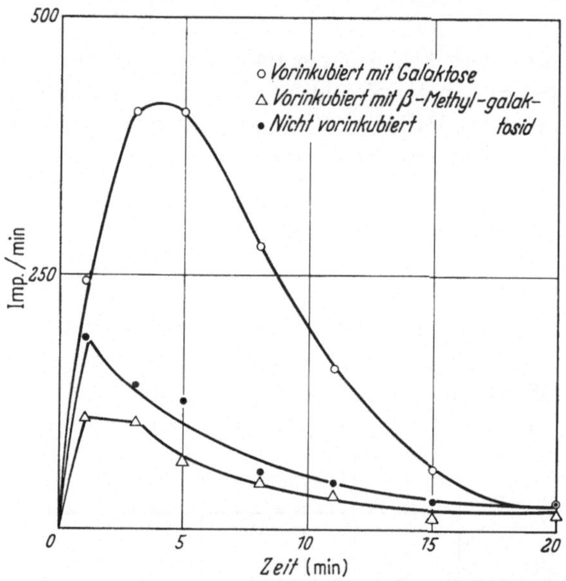

Abb. 8. Wirkung der Vorsättigung von Bakterienzellen (E.coli) mit Galaktose auf die Eintrittsgeschwindigkeit von ^{14}C-Galaktose in Gegenwart von Dinitrophenol. Ruhende Zellsuspensionen (196 μg pro ml) wurden 15 min bei 25° präinkubiert mit $5 \cdot 10^{-5}$ M ^{12}C-Galaktose oder $5 \cdot 10^{-4}$ M β-Methylgalaktosid. Kontrollzellen wurden ohne Zusätze präinkubiert. Nach dem Zentrifugieren wurden die Zellen resuspendiert im gleichen Volumen eines frischen Mediums, das 10^{-3} M Dinitrophenol und $5 \cdot 10^{-5}$ M ^{14}C-Galaktose ($2,2 \cdot 10^{5}$ Impulse pro Minute pro Mikromol) enthielt. (HORECKER et al. 1961)

An Bakterien zeigt die in Abb. 8 wiedergegebene Beobachtung von HORECKER et al. im (ausgleichenden System) des Galaktoseaustritts aus E. coli einen Gegentransport, der, in Übereinstimmung mit anderen Hinweisen, dieses System als Trägersystem kennzeichnet (im Gegensatz zu der ursprünglichen Deutung im Sinne freier Diffusion). An Zellen, die nach Sättigung mit ^{12}C-Galaktose in einer DNPhaltigen, gleichkonzentrierten, ^{14}C-Galaktoselösung suspendiert werden, entsteht ein vorübergehender, sich nach kurzer Zeit umkehrender, intensiver Einwärts-Strom von

^{14}C-Galaktose, der bei den Kontrollzellen (nicht mit „kalter“ Galaktose vorgesättigt) fehlt.

Neben dem Nachweis der Bewegung des Trägersubstratkomplexes erlaubt der Gegentransport noch eine zweite Entscheidung, nämlich ob zwei Substrate, die sich an einem Transportsystem wechselseitig hemmen, gemeinsame Affinität zum Träger oder nur gemeinsame Affinität zu einem Enzym oder anderen Bestandteilen eines komplexeren Systems besitzen. Nur bei gemeinsamer Affinität zum Träger ist der Gegentransport zu erwarten. Ein gemeinsamer Träger ergibt sich aus solchen Beobachtungen daher nicht nur für die Aldosen, Xylose, Mannose und Glucose am Erythrocyten[37,47], sondern auch für die Ketose, Sorbose und die Aldose Glucose und Galaktose an Hefezellen[2,6].

3. Beobachtungen an Permease-Systemen

Mit dem Gegentransport nahe verwandt, bzw. durch ihn bedingt, ist eine weitere Erscheinung bei gleichzeitiger Anwesenheit von zwei Substraten für das gleiche System: die Transportbeschleunigung für ein erstes Substrat bei Anwesenheit eines zweiten in niedriger, auf beiden Membranseiten gleicher, Konzentration[64].

Bei hohen Konzentrationen führt die Anwesenheit dieses zweiten Substrates, wie zu erwarten, zu kompetitiver Hemmung. Das quantitative Ausmaß dieser Hemmung läßt sich berechnen aus Beziehungen entsprechend der Gl. (1), in denen in die beiden Sättigungsterme die Konzentration I und die Michaelis-Konstante K_I des zweiten Substrats in geeigneter Weise eingesetzt wird. Dabei zeigt sich, daß die Hemmung negatives Vorzeichen annimmt (d. h. in Beschleunigung des Transports übergeht), wenn

$$I/K_I < S_1 S_2/K_m{}^2 . \tag{6}$$

Abb. 9 zeigt die zu erwartende Abhängigkeit von den Konzentrationen, berechnet für ein konkretes Beispiel.

Diese paradoxe Beschleunigung durch einen kompetitiven Inhibitor ist bisher an Ausgleichssystemen nicht nachgewiesen worden*, dagegen bei Bergauftransport an Bakterien.

---

* *Anmerkung bei der Korrektur:* Inzwischen ist in unserem Laboratorium dieser Nachweis für den Glucoseaustritt aus Menschenerythrocyten in Gegenwart von Mannose geführt worden.

Abb. 10 zeigt eine Beobachtung von KEPES[27], bei der an E. coli der Eintritt von Thiodigalaktosid (TDG) durch Thiomethylgalaktosid (TMG) in den Konzentrationen 5×10^{-4} M und 10^{-4} M gehemmt, in der Konzentration 2×10^{-5} M dagegen beschleunigt wird.

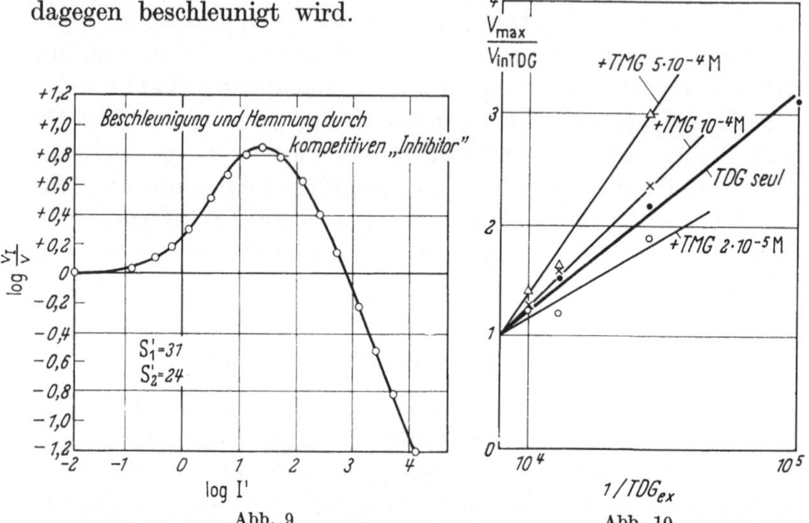

Abb. 9 Abb. 10

Abb. 9. Beschleunigung eines Trägertransports eines Substrats S in Gegenwart eines zweiten Substrats I in gleicher Konzentration auf den beiden Seiten der Membran. S_1' und S_2' sowie I' sind „relative" Konzentrationen (d. h. die Quotienten aus den absoluten Konzentrationen und den zugehörigen Michaeliskonstanten). Die Wirkung von I auf den Transport von S wird angegeben durch das Verhältnis $\frac{v_I}{v}$, wobei v_I die Transportgeschwindigkeit von S in Gegenwart von I, v die gleiche Transportgeschwindigkeit in Abwesenheit von I darstellt

Abb. 10. Beschleunigende Wirkung eines kompetitiven Transportsubstrates TMG auf die Eintrittsgeschwindigkeit des Substrates TDG in Bakterienzellen. Lineweaver-Burksche Darstellung: Ordinate reziproke relative Eintrittsgeschwindigkeit von TDG, Abszisse reziproke Außenkonzentration von TDG. TMG = Thiomethylgalaktosid, TDG = Thiodigalaktosid. Die Konzentrationen 10^{-4} M und $5 \cdot 10^{-4}$ M von TMG hemmen, die Konzentration $2 \cdot 10^{-6}$ M beschleunigt den Eintritt von TDG (KEPES 1960)

An (ebenfalls bergauftransportierenden) Aminosäure-Systemen liegen ähnliche Beobachtungen vor, sowohl an Bakterien[8] als am Darm[57,65,66].

Auch eine weitere Beobachtung aus dem Gebiet der Permeasen deutet möglicherweise auf ein Trägerelement im Eintritts-System. Charakteristisch für die Permeasen-Systeme ist die Möglichkeit, das aus sehr niedrigen Konzentrationen akkumulierte markierte Substrat durch Zusatz hoher Konzentrationen nicht markierten Substrats wieder aus den Zellen „auszutreiben". Ob dieser Effekt ausreichend durch die kompetitive Hemmung des Eintrittssystems (bzw. durch die Abnahme des Akkumulationsverhältnisses mit steigender Substrat-Konzentration) zu deuten ist oder ob eine Komponente daran beteiligt ist, die auf Gegentransport beruht, ist wohl gegenwärtig nicht mit Sicherheit zu entscheiden. Der von HORECKER et al. erhobene Befund (Abb. 11), daß Zusatz von Dinitrophenol, wie erwähnt, qualitativ ähnlich wirkt wie Zusatz

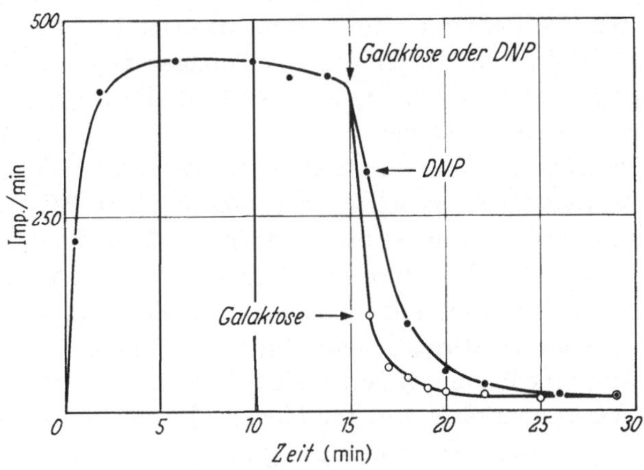

Abb. 11. Austreibung radioaktiver Galaktose durch „kalte“ Galaktose bzw. durch Dinitrophenol aus Bakterienzellen (E.coli). Zu ruhenden Zellsuspensionen, die bei 25° mit $5 \cdot 10^{-5}$M ^{14}C-Galaktose 15 min inkubiert waren, wurde beim Pfeil entweder 1,0 M ^{12}C-Galaktose oder 0,1 M Dinitrophenol in solchen Mengen zugesetzt, daß Konzentrationen von $5 \cdot 10^{-3}$ M Galaktose und 10^{-3} M Dinitrophenol entstehen. Die Austreibungsgeschwindigkeit beim Galaktosezusatz ist größer als bei Dinitrophenolzusatz, vermutlich auf Grund von Gegentransport (HORECKER et al. 1961)

hoher Substratkonzentrationen, aber zu deutlich langsamerem Austritt führt, könnte jedoch darauf beruhen, daß beim Dinitrophenolzusatz das Element des Gegentransports fehlt.

Die Deutung der Permeasen-Transporte im Sinne eines Trägertransports ist ein gemeinsames Element in den Diskussionen der Systeme durch MONOD[8], KEPES[27] und HORECKER[25, 26]. Sie wirft die Frage auf, ob die Permease selbst als der Träger zu betrachten ist, oder im Sinne der Abb. 2 als ein Enzym, das die Trägersubstratreaktion katalysiert.

HORECKER hat gezeigt, daß im Galaktosesystem die Freisetzung markierten Substrats aus den Zellen nicht nur durch Zusatz hoher Konzentrationen von Galaktose möglich ist, sondern auch durch hohe Glucosekonzentrationen. Da andererseits die Eintrittsschwindigkeit von Galaktose durch Zusatz von Glucose in vergleichbarer Konzentration nicht gehemmt wird, diskutiert Horecker die Möglichkeit, daß die Permease, d. h. das Element mit hoher Strukturspezifität, nur Affinität zur Galaktose, nicht dagegen zu Glucose besitzt und daß das Element, dessen gemeinsame Affinität zu Glucose und Galaktose die Grundlage der „Austreibungs"-Reaktion der beiden Zucker bildet, das Trägermolekül ist. Für eine solche Deutung kann auch angeführt werden, daß die Michaelis-Konstanten der Permeasen im allgemeinen um 1—2 Größenordnungen tiefer liegen als diejenigen von Trägersystemen an anderen Zellen, (vor allem am Erythrocyten und am Darm) und daß sie darin Enzymen näher stehen.

Die oben behandelten quantitativen Konsequenzen des Trägerprinzips wurden für Ausgleichssysteme abgeleitet. Die Wahrscheinlichkeit einer Trägernatur der bergauf transportierenden Permeasen (insbesondere nahegelegt durch die Hinweise auf Gegentransport) wirft daher die Frage auf, unter welchen zusätzlichen Voraussetzungen Trägersysteme bergauf transportieren können und ob die nötigen Zusatzbedingungen derart sind, daß die oben abgeleiteten Konsequenzen qualitativ erhalten bleiben. In diesem Fall würden die hier vorgeschlagenen Deutungen gerechtfertigt.

Einige (gegenwärtig noch fragmentarische) Möglichkeiten in dieser Richtung sollen im folgenden Abschnitt behandelt werden.

4. Der Trägermechanismus beim Bergauftransport

Möglichkeiten, aus den bisherigen, nur Ausgleich bewirkenden, Systemen solche zu machen, die bergauf zu transportieren vermögen, lassen sich gut an einer Beziehung demonstrieren, die die Geschwindigkeit des Ausgleichstransports in ihrer Abhängigkeit von den Parametern des Transportsystems in formal etwas anderer Weise darstellt als Gl. (1), mit dieser Gleichung aber materiell identisch ist:

$$V = D'\left(\frac{C_1 S_1}{K_m} - \frac{C_2 S_2}{K_m}\right). \qquad (7)$$

Aus dieser Beziehung ergibt sich, daß die Transportgeschwindigkeit Null ist, wenn die drei Größen C, S und K_m auf beiden Seiten der Membran gleich sind. Gleichung (2) zeigt unmittelbar, daß für $S_1 = S_2$ die Geschwindigkeit Null ist (Ausgleichstransport). Daraus ergibt sich für diesen Fall, daß auch $C_1 = C_2$ sein muß.

Wird nun aber, unter Beibehaltung von $S_1 = S_2$, die Symmetrie bezüglich C oder K_m gestört, so muß das zu einem Bergauftransport führen.

In bezug auf C wurde eine Möglichkeit einer Asymmetrie ($C_1 \neq C_2$) bereits erwähnt: der Gegentransport. Wie Abb. 3 zeigt, wird diese Asymmetrie durch den Zusatz des zweiten Substrates R erzeugt und führt zu einem Bergauftransport. Eine andere Möglichkeit wäre der Zusatz von C auf einer Membranseite, beispielsweise durch metabolische Reaktionen im Innern einer Zelle. Einen Modellversuch dafür haben (allerdings nicht für Transport von Zuckern, sondern von Aminosäuren), CHRISTENSEN und OXENDER[5] gegeben: sie zeigten, daß das (von CHRISTENSEN 1955 als möglicher Träger diskutierte) Pyridoxal bei einseitigem Zusatz zu einem symmetrischen System einen Bergauftransport von Aminosäure bewirkt. Auch einseitiger Kaliumzusatz wirkte gleich, in Übereinstimmung mit CHRISTENSENs Annahme, daß der Komplex auch Kalium enthält.

Die zweite Möglichkeit einer Asymmetrie, nämlich $K_{m1} \neq K_{m2}$ ist vorstellbar, wenn auf einer Membranseite metabolische Reaktionen mit dem Träger C dessen Affinität zum Substrat verändern. Aus thermodynamischen Gründen ist das nur mit entsprechenden Energieumsetzungen möglich, so daß hier eine weitere Möglichkeit liegt, die freie Energie von Stoffwechselvorgängen in der Zelle unmittelbar für Bergauftransporte verwertbar zu machen. Beispiele solcher Systeme sind die Mechanismen, die von SOLOMON[49] und von SHAW[48] für den Bergauftransport von Natrium und von Kalium durch die Erythrocytenmembran in Vorschlag gebracht worden und von anderen Autoren[20, 24] mehrfach benützt worden sind.

Für einen Zuckertransport ist ein experimenteller Hinweis in dieser Richtung der Befund von HORECKER et al.[25], daß am Galaktose-Permease-System der numerische Wert von K_m durch Dinitrophenol deutlich erhöht wird. Die Hemmwirkung von DNP auf den Eintritt von Galaktose (vermutlich infolge des Fehlens von energiereichem Phosphat) könnte danach so zu deuten sein, daß ein normalerweise bestehender Unterschied zwischen K_{m1} und K_{m2} (vermutlich bedingt durch Reaktionen des Trägers mit energiereichem Phosphat) unter diesen Umständen aufgehoben wird.

Ob und in welcher Weise solche Möglichkeiten bei den transcellulären Bergauftransporten verwirklicht sind, die mit Zuckern beobachtet werden, ist unbekannt. Eine Hypothese, die vor kurzem von CRANE[11] für die Zuckerresorption im Darm geäußert worden ist, arbeitet jedoch mit der angenommenen Asymmetrie in bezug auf C und sei deshalb erwähnt.

CRANE geht aus von der in mehreren Laboratorien gefundenen Beziehung zwischen Zuckerresorption und den Konzentrationen der Kalium- und Natriumionen. In Abwesenheit von Natriumionen ist kein Bergauftransport von Zucker durch die Epithelzellschicht der Darmwand möglich[12, 42]. Herzglykoside, deren Hemmwirkung auf Kationentransporte von anderen Zellen her bekannt ist[21], verhindern den Bergauftransport von Zucker auch in Anwesenheit von Natriumionen. Neue Befunde von KLEINZELLER und KOTYK* zeigen die gleichen Eigentümlichkeiten für die Galaktoseakkumulation in Nierenschnitten.

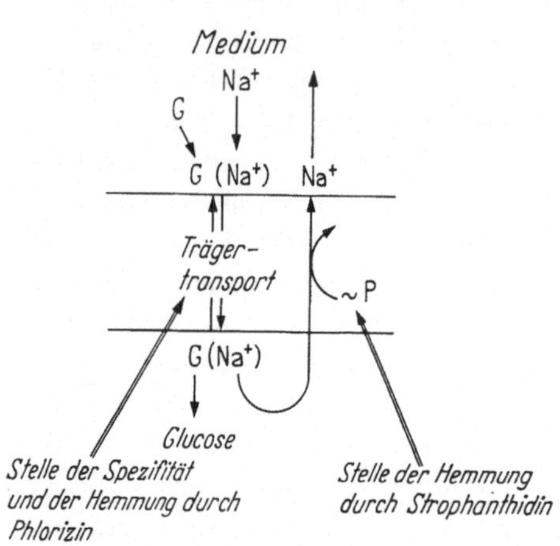

Abb. 12. Vorstellung von CRANE et al. zur Deutung der Abhängigkeit der Zuckerresorption im Darm von der Anwesenheit von Natrium und von der Tätigkeit der Natriumpumpe. Das Glucosemolekül G bildet mit Natriumionen und einem Träger einen Komplex, der die Membran passiert und auf der Innenseite Glucose und Natrium freisetzt. Die Entfernung des Natriums durch die Natriumpumpe beeinflußt den Gradienten des Transportkomplexes und führt dadurch zur Akkumulation von Glucose in der Zelle (CRANE et al. 1961)

CRANE schlägt nun für diese Befunde die folgende Deutung vor. Der Transportkomplex für den Eintritt von Zucker in die Epithelzelle enthält außer dem Träger noch Natrium, das, gemeinsam mit dem Zuckermolekül, im Innern der Zelle freigesetzt wird. Solange die Natriumpumpe Natriumionen aus der Epithelzelle entfernt, wird dadurch (analog zu dem erwähnten Modellversuch von CHRISTENSEN und OXENDER) ein Gradient für den Trägerkomplex aufrechterhalten, selbst wenn die Zuckerkonzentrationen auf beiden Seiten der Membran gleich sind. Der unmittelbare Energielieferant für den Bergauftransport des Zuckers

* Persönliche Mitteilung.

wäre in diesem Fall die Natriumpumpe. Abb. 12 zeigt den vorgeschlagenen Mechanismus schematisch.

Zweck dieser Erörterung war jedoch nicht in erster Linie die Anwendung auf die Bergauftransporte in Niere und Darm, die noch im Stadium erster Hypothesen steht, sondern der Hinweis, daß im Rahmen der diskutierten Möglichkeiten (auch bei Bergauftransporten) Beziehungen zwischen Transportgeschwindigkeit und Substratkonzentrationen zu erwarten sind, die formal der Gl. (1) weitgehend ähneln, so daß, mutatis mutandis, auch für solche Systeme qualitativ ähnliche kinetische Forderungen sich ergeben werden.

5. Insulinwirkungen

Einer der ersten Befunde, die auf einen Angriff des Insulins an der Zuckerpassage durch die Zellmembran hinwies, wurde 1939 von LUNDSGAARD mitgeteilt[33]. Er fand, daß Insulin die Zuckeraufnahme im perfundierten quergestreiften Muskel erhöht, obwohl im Inneren die Konzentration der freien Glucose praktisch Null ist. Unter diesen Umständen, aus denen sich eine limitierende Funktion des Zuckereintritts in die Zelle ergibt, wird eine direkte Wirkung des Insulins auf die Membranpassage sehr wahrscheinlich.

Neue Impulse enthielt diese Interpretationsrichtung durch die Untersuchungen von LEVINE[32] über den Einfluß des Insulins auf die Zuckerverteilung im Organismus eviszerierter Hunde. LEVINE studierte die Verteilung intravenös injizierter Galaktose (eines Zuckers der nicht oder nur langsam verwertet wird). Während Galaktose sich normalerweise nur auf das extracelluläre Wasser verteilt, fand er unter dem Einfluß von Insulin ein Verteilungsvolumen, das dem Gesamtwasser nahekommt. Er nahm daher an, daß Insulin Muskelzellmembranen, die normalerweise für Galaktose undurchlässig sind, durchlässig macht. LEVINEs Befunde wurden bestätigt und erweitert durch DRURY und WICK[15] sowie durch verschiedene Untersuchungen der Arbeitsgruppe um PARK[37,38,39].

Ein besonders geeignetes Objekt für die Analyse von Zuckertransporten durch Zellmembranen wurde von R. B. FISHER[18] im perfundierten Säugetierherzmuskel gefunden. An diesem Objekt wurden Untersuchungen sowohl mit Glucose als auch mit nichtmetabolisierten Zuckern, vor allem L-Arabinose, durchgeführt. Für beide Zucker ergab sich, daß der Durchtritt durch die Membran beschleunigt wird, im Falle der L-Arabinose in beiden Richtungen[17, 38] (s. Abb. 13).

In neueren Analysen an diesem Objekt wurde versucht, die Insulinwirkung quantitativ zu deuten im Sinne einer Beeinflussung eines Trägersystems, wie es durch die Gl. (1) dargestellt wird, und sie durch die numerische Veränderung der beiden Parameter V_{max} und K_m zu charakterisieren. Post u. Mitarb.[39] führten diese Analyse mit Glucose durch, wobei sowohl die Parameter des Phosphorylierungssystems als diejenigen des Transportsystems bestimmt wurden. R. B. Fisher arbeitete mit den nicht vergärbaren Zuckern D-Xylose und L-Arabinose, was die Analyse etwas vereinfacht, da nur die Parameter des Transportsystems in Betracht zu ziehen sind.

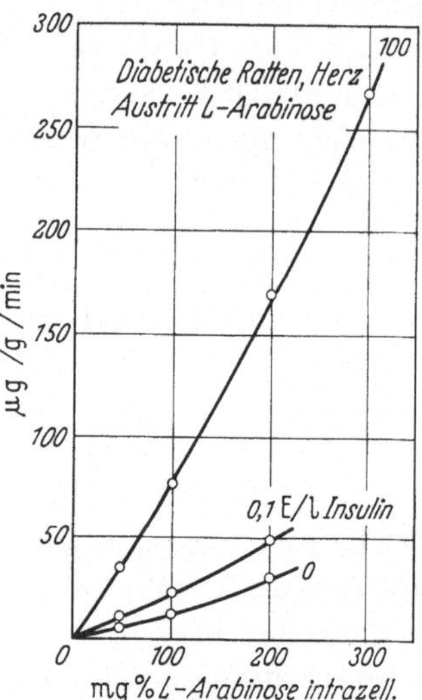

Abb. 13. Wirkung von Insulin auf die Austrittsgeschwindigkeit von L-Arabinose aus dem perfundierten Herzmuskel diabetischer Ratten. (Nach Post, Morgan und Park 1961)

Die Ergebnisse der beiden Untersuchungsreihen sind in Tab. 2 zusammengestellt. Gemeinsam ist in beiden Versuchsreihen eine Erhöhung der Werte von K_m, d. h. eine Herabsetzung der Affinität des untersuchten Zuckers zum Transportsystem. Dagegen unterscheiden sie sich in der Wirkung des Insulins auf den Kapazitätsfaktor: Während in den Versuchen von Post u. Mitarb. die Maximalgeschwindigkeit erhöht wird, wird sie in den Versuchen von Fisher herabgesetzt. Fisher deutet daher die Beschleunigung des Zuckerdurchtritts durch die Membran im Sinne der oben dargestellten Beziehung zwischen Affinität und Transportgeschwindigkeit bei hoher Sättigung: Erhöhung der Geschwindigkeit bei Verminderung der Affinität. Die dafür notwendige Voraussetzung, daß die Substratkonzentration erheblich höher ist als die Michaelis-Konstante, ist in seinen Versuchen, die mit Zuckerkonzentrationen von 30 mM durchgeführt worden sind, erfüllt.

Neben der Diskrepanz in bezug auf die Insulinwirkung auf den Kapazitätsfaktor fällt auch auf, daß die von Post u. Mitarb. für

Tabelle 2. *Wirkung von Insulin auf die Parameter des Zuckertransportsystems im Herzmuskel*

G = D-Glucose (Versuche von Morgan, Post und Park 1961)
A = L-Arabinose, X = D-Xylose (Versuche von R. B. Fisher 1961)

	K_m (mM)			v_{max} (Einheiten ungleich)		
	G	A	X	G	A	X
ohne Insulin . . .	8,7	0,062	0,21	18,4	11	8
mit Insulin . . .	27,6	25	6,6	90,0	2	3,5

Glucose gefundenen Werte von K_m beträchtlich höher sind als die von Fisher für die nichtmetabolisierten Pentosen gefundenen Werte. Vielleicht sind diese Diskrepanzen ein Hinweis darauf, daß die Voraussetzungen, unter denen die Gln. (1) und (1a) sowie ihre verschiedenen Folgerungen abgeleitet worden sind (z. B. die Annahme des Reaktionsgleichgewichtes), Vereinfachungen darstellen, die nur unter beschränkten Bedingungen zu widerspruchsfreien

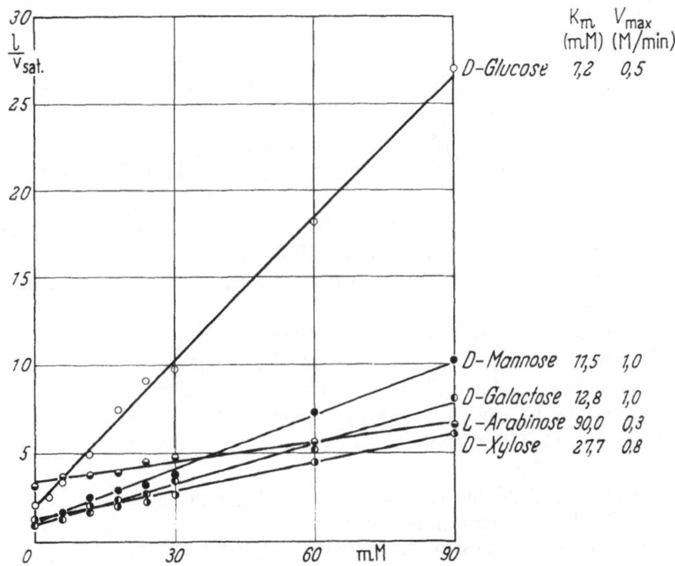

Abb. 14. Graphische Bestimmung der Parameter K_m und v_{max} für fünf Zuckerarten aus Austrittsversuchen bei 37° an Menschenerythrocyten. Entgegen der Erwartung unterscheiden sich die Zucker nicht nur in den ermittelten Werten von K_m, sondern auch in v_{max} (Wilbrandt 1961)

Resultaten führen, und daß die Frage, was mit den beiden Parametern des Trägersystems wirklich erfaßt wird, einer verfeinerten Analyse bedarf.

Ähnliche Schlüsse wurden früher[58] aus den in Abb. 14 wiedergegebenen Befunden gezogen. Sie zeigen, daß die experimentelle Ermittlung der beiden Transportparameter für verschiedene Zucker an Menschenerythrocyten nicht, wie zu erwarten, nur Unterschiede in K_m ergibt, sondern auch in ausgesprochener Weise in V_{max}. Letzteres wäre bei Gültigkeit der Gl. (1a) und ihrer Voraussetzungen ($V_{max} = D' C_t$) höchstens in geringfügigem Maß zu erwarten, wenn alle Zucker den gleichen Träger benützen.

Literatur

1 BARANY, E., and E. SPERBER: Scand. Arch. Physiol. **81**, 290—299 (1939).

2 BURGER, M., L. HEIMOVA and A. KLEINZELLER: Biochem. J. **71**, 233—242 (1959).

3 CAHILL, jr., G. F., J. ASHMORE, A. S. EARLE and S. ZOTTU: Am. J. Physiol. **192**, 491—496 (1958).

4 CHRISTENSEN, H. N.: Science **122**, 1087—1088 (1955).

5 CHRISTENSEN, H. N., and D. L. OXENDER: Amer. J. clin. Nutr. **8**, 131 to 136 (1960).

6 CIRILLO, V. P.: Bact. Proc. **23**, 108—109 (1959).

7 COHEN, G. N., and J. MONOD: Bact. Rev. **21**, 169—194 (1957).

8 COHEN, G. N., et H. V. RICKENBERG: Ann. Inst. Pasteur. **91**, 693—720 (1956).

9 CRANE, R. K.: Physiol. Rev., in press (1961).

10 CRANE, R. K., R. A. FIELD and C. F. CORI: J. biol. Chem. **224**, 649—662 (1957).

11 CRANE, R. K., D. MILLER and I. BIHLER: Symp. Prag, Publ. House of the Czechoslovak Acad. Sci. in press (1961).

12 CSAKY, T. Z., and L. ZOLLIKOFER: Amer. J. Physiol. **198**, 1056—1058 (1960).

13 DETTLING, P.: Dissertation, Bern 1954.

14 DOUDOROFF, M., W. Z. HASSID, E. W. PUTHAM, A. L. POTTER and J. LEDERBERG: J. biol. Chem. **179**, 921—923 (1949).

15 DRURY, D. R., and A. N. WICK: Amer. J. Physiol. **171**, 721 (1952).

16 EGE, R., E. GOTTLIEB and N. W. PAKESTRAW: Amer. J. Physiol. **72**, 76—83 (1925).

17 FISHER, R. B.: Ciba Found. Symp. on Enzymes and Drug Action (1961) Churchill Ltd. London im Druck.

18 FISHER, R. B., and D. B. LINDSAY: J. Physiol. (Lond.) **131**, 526—541 (1956).

19 FRIDHANDLER, L., and J. H. QUASTEL: Arch. Biochim. **56**, 412—423 (1955).

20 Glynn, I. M.: Progr. Biophys. 8, 241—307 (1957).
21 Hajdu, S., u. E. Leonard: Pharmacol. Rev. 11, 173—209 (1959).
22 Helmreich, E., and C. F. Cori: J. biol. Chem. 224, 663—679 (1957).
23 Helmreich, E., and H. N. Eisen: J. biol. Chem. 234, 1958—1965 (1959).
24 Hodgkin, A. L.: Proc. roy. Soc. B. 148, 1—37 (1957).
25 Horecker, B. L., M. J. Osborn, W. L. McLellan, G. Avigard and C. Asensio: Symp. on Membrane Transport and Metabolism Prag, Publ. House of the Czechoslovak Acad. Sci. in press (1961).
26 Horecker, B. L., J. Thomas and J. Monod: J. biol. Chem. 235, 1586 to 1590 (1956).
27 Kepes, A.: Biochim. biophys. Acta 40, 70—84 (1960).
28 Kozawa, S.: Biochem. Z. 60, 231—256 (1914).
29 Lefèvre, P. G.: J. gen. Physiol. 31, 505—527 (1948).
30 Lefèvre, G. P.: Symp. Soc. exp. Biol. 8, 118—135 (1954).
31 Lefèvre, P. G., and J. K. Marshall: Amer. J. Physiol. 194, 333—337 (1958).
32 Levine, R., and M. S. Goldstein: Recent Progr. Hormone Res. 11, 343 to 375 (1955).
33 Lundsgaard, E.: Upsala Läk.-Fören. Förh. 45. 143—151 (1939/1940).
34 Meldahl, K. P., u. S. L. Ørskov: Skand. Arch. Physiol. 83, 266—280 (1940).
35 Nirenberg, W., and J. F. Hogg: J. Amer. chem. Soc. 78, 6210—6212 (1956).
36 Nirenberg, M. W., and J. F. Hogg: J. Amer. chem. Soc. 80, 4407—4412 (1958).
37 Park, C. R., R. L. Post, C. F. Kalman, J. H. Wright, jr., L. H. Johnson and H. E. Morgan: Ciba Colloquia Endocrin. 9, 240—260 (1956).
38 Park, C. R., D. Reinwein, M. J. Henderson, E. Cadenas and H. E. Morgan: Amer. J. Med. 26, 647—684 (1959).
39 Post, R. L., H. E. Morgan and C. R. Park: J. Biol. Chem. im Druck (1961).
40 Randle, P. J., u. G. H. Smith: Biochem. J. 70, 490—500 (1958).
41 Randle, P. J., u. G. H. Smith: Biochem. J. 70, 501—508 (1958).
42 Riklis, E., and J. H. Quastel: Canad. J. Biochem. 36, 347—362 (1958).
43 Rosenberg, Th.: Acta chem. scand. 2, 14—33 (1948).
44 Rosenberg, Th.: Biol. et Path. in press (1961).
45 Rosenberg, Th., u. W. Wilbrandt: Int. Rev. Cytol. 1, 65—92 (1952).
46 Rosenberg, Th., and W. Wilbrandt: Exp. Cell. Res. 9, 49—67 (1955).
47 Rosenberg, Th., and W. Wilbrandt: J. gen. Physiol. 41, 289—296 (1957).
48 Shaw, T. I.: Ph. D. Thesis, Cambridge University, England 1954.
49 Solomon, A. K.: J. gen. Physiol. 36, 57—110 (1952/53).
50 Widdas, W. F.: J. Physiol. (Lond.) 118, 23—39 (1952).
51 Widdas, W. F.: J. Physiol. (Lond.) 125, 163—180 (1954).
52 Wilbrandt, W.: Pflügers Arch. ges. Physiol. 241, 302 (1938).
53 Wilbrandt, W.: Symp. Soc. exp. Biol. 8, 136 (1954).
54 Wilbrandt, W.: J. cell. comp. Physiol. 47, 137—145 (1956).
55 Wilbrandt, W.: Dtsch. med. Wschr. 82, 1153—1158 (1957).

[56] WILBRANDT, W.: Mod. Probl. Pädiat. 4, 30—49 (1959).
[57] WILBRANDT, W.: Ciba Found. Coll. im Druck (1961).
[58] WILBRANDT, W.: Symp. on transport and metabolis,. Prag 1961, i. Druck.
[59] WILBRANDT, W., S. FREI and T. ROSENBERG: Exp. Cell. Res. 11, 59—66 (1956).
[60] WILBRANDT, W., H. LAUENER u. E. GÜNSBERG: Helv. physiol. pharmacol Acta 5, C20—C22 (1947).
[61] WILBRANDT, W., u. T. ROSENBERG: Helv. physiol. pharmacol.Acta 8, C82 bis C83 (1950).
[62] WILBRANDT, W., and TH. ROSENBERG: Helv. physiol. pharmacol. Acta 9, C86—C87 (1951).
[63] WILBRANDT, W., and TH. ROSENBERG: Exp. Cell. Res. i. Druck (1961).
[64] WILBRANDT, W., and TH. ROSENBERG: Pharmacol. Rev. i. Druck (1961).
[65] WISEMAN, G.: J. Physiol. (Lond.) **120**, 63—72 (1953).
[66] WISEMAN, G.: J. Physiol. (Lond.) **133**, 626—630 (1956).

Diskussion

Diskussionsleiter: BÜCHER, *Marburg*

KLEINZELLER (Prag): Wir haben in Nierenschnitten gefunden (KLEINZELLER und KOTYK), daß die Galaktoseakkumulation, von CRANE früher beschrieben, Na-abhängig ist. Wenn Na^+ durch Li ersetzt wird, kommt es nur zu einem Ausgleich der Galaktosekonzentration, und Ouabain blockiert auch anaerob die Galaktosepermeierung. Das Transportsystem für Galaktose in der Nierenrinde hat also dieselben Eigenschaften wie das der Darmschleimhaut (CRANE u. Mitarb.). Dr. KÉPÈS sagte mir gestern, daß das Permease-System nicht von den Elektrolyten abhängig ist. Wir hätten dann zwei verschiedene aktive Systeme, von denen nur eines elektrolyt-abhängig ist. Professor WILBRANDT hat in Prag erwähnt, daß der Eintritt von Monosacchariden in den Erythrocyten elektrolytabhängig ist; wir finden aber in Hefezellen, daß hier der Ausgleich nicht von Elektrolyten beeinflußt ist. Soll man nun annehmen, daß es also vier verschiedene Zuckertransportsysteme gibt, von denen zwei elektrolytabhängig sind, zwei andere nicht?

WILBRANDT: Wir können im Augenblick die Frage, ob bei den Erythrocyten der Zuckertransport vom Natrium abhängig ist, nicht mit Sicherheit beantworten. Vorerst kann man mit der Möglichkeit rechnen, daß dieser Natriumeinfluß beschränkt ist auf die akkumulierenden Systeme. Weitere Untersuchungen sind nötig.

FISCHER (Frankfurt): Wir haben folgendes Problem, das aus der Praxis kommt. Wenn man Zellen mit Steroiden versetzt, und zwar denke ich in erster Linie an Prednisolon, und zwar in Konzentrationen, die pharmakologisch Usus sind, etwa 0,1 μmol bis 1 μmol pro cm^3, dann kann man einige Transportphänomene beobachten. Derartige Zellen, jetzt spreche ich von Erythrocyten, nehmen vermindert Glucose auf, können auch nicht mehr gut Inosin aufnehmen, und es sind auch Abnormitäten im Phosphatstoffwechsel. Bei Konzentrationen von 0,1 μmol pro cm^3 ist die Inkorporation von ^{32}P in die verschiedensten organischen Phosphatfraktionen verhindert oder gehemmt, am deutlichsten in ATP. Wird die Konzentration von Prednisolon gesteigert, sehr hoch gesteigert, so ist auch der Einbau von Radiophosphat, von anorganischem Phosphat in die Zelle gehemmt. Also Phosphat kann nicht mehr in die Zelle hinein. Ein weiteres Phänomen: Derartige Erythrocyten sind mechanisch resistenter. Diese mechanische Resistenz hält bei einem Anfangszusatz zu Blutkonserven auf über 50 Tage Beobachtungsdauer an, so daß man daran denken könnte, es handele sich hierbei um eine ziemlich unspezifische Beeinflussung der Grenzmembran, die zu einer mechanischen Verfestigung führen kann. Ich möchte noch sagen, daß dieser Befund nicht nur für die Erythrocyten gültig ist, sondern auch für andere Zellen. So ist ja bekannt, daß z. B. Leukocyten, mit höheren Dosen Prednisolon versetzt, nicht mehr phagocytieren können; wir haben also auch hier eine Beeinflussung der Permeabilität. Meine Frage ist nun: hat einer der anwesenden Spezialisten auch schon Steroide dieser Art in ihrer Wirkung auf den Transport, auf verschiedenartige Transportphänomene untersucht?

Wilbrandt: Wir haben beim Glucosetransport an Erythrocyten mit Desoxycorticosteron Hemmeffekte gesehen [T. Rosenberg u. W. Wilbrandt, Helv. physiol. pharmacol. Acta **15**, 168—176 (1957)], allerdings mit sehr hohen Konzentrationen (10^{-3} g pro ml in Substanz zugesetzt, d. h. gesättigte Lösungen mit Boden-Körper). Von verschiedenen Seiten ist auf Ähnlichkeiten zwischen Steroidwirkungen und Phlorrhizinwirkungen hingewiesen worden. Das Beispiel, daß Sie erwähnen, läßt mich an diese Beziehungen denken, weil das Phlorrhizin, das ja vor allem Zuckertransporte hemmt, von Harris auch als Inhibitor der Phosphataufnahme beschrieben worden ist.

Es wäre also ganz interessant zu sehen, ob die Effekte, die Sie mit Steroiden bekommen, mit Phlorrhizin reproduzierbar sind. Dann würde sich diese Parallele hier noch etwas verbreitern. Für ihre Deutung ist darauf hingewiesen worden, daß die Molekülstruktur des Phorrhinzins bei Ringschluß zwischen der Carbonylgruppe und einer Hydroxylgruppe des benachbarten Rings (z. B. durch eine Wasserstoffbrücke) in einer steroidähnlichen Form geschrieben werden kann, in etwas ähnlicher Weise wie im Fall des Stilboestrols.

Jatzkewitz (München): Ich glaube nicht, daß die Steroidhormonwirkung auf den aktiven Transport beim Erythrocyten spezifisch ist. Wenn man annimmt, daß sich Hormonwirkung und -transport an der Membran manifestieren, dann scheint mir für die Erklärung der spezifischen Hormonwirkung eher die spezifische Membran des Erfolgsorgans nötig zu sein.

Hoffmann-Berling (Heidelberg): Ich möchte Herrn Képès zum Wesen der Permease eine genetische Frage stellen. In der Arbeit von Horecker, Thomas und Monod, die herangezogen wurde, schließen die Autoren aus ihren kinetischen Daten, daß die Permease für Galaktose ein mehrteiliges System ist und aus mindestens zwei Enzymen besteht: E_1 und E_2, von denen E_1 den Eintritt der Galaktose in die Membran katalysiert und vielleicht die Galaktose mit einem aktivierten "carrier" koppelt und E_2 den Austritt aus der Membran, also offenbar die Spaltung dieses Carrier-Galaktose-Komplexes, bewirkt. Diese Vorstellung haben Sie ja schon früher entwickelt, und die erinnert ja deutlich an dieses Hokinsche Schema über den Umsatz des Diglycerides, das wir heute morgen gehört haben. Wenn nun die Permease aus zwei getrennten Enzymen besteht, dann sollte genetisch folgender Versuch möglich sein: Man sollte ausgehen von zwei permease-negativen Mutanten, von denen die eine den Locus für das Enzym E_1 in ihrem negativen Allel trägt und den Locus E_2 im positiven Allel und die andere die komplementäre genetische Konstitution hätte. Durch Kreuzung sollte man eine diploide Zelle herstellen können, in der sich die beiden positiven Gene zur Funktion komplettieren. Man sollte also eine Zelle bekommen, die permease-positiv ist, weil ja jedes der beiden Allele mindestens einmal in seiner positiven Form vorliegt. Wenn dagegen die Permease ein einheitliches Enzym ist, dann sollte diese funktionelle Komplettierung nicht möglich sein. Sie kommen aus dem Institut Pasteur, das eine hervorragende genetische Abteilung hat, und ich möchte daher fragen: Sind diese sehr naheliegenden Versuche bisher angestellt worden und können Sie etwas darüber sagen?

KEPES: My kinetic studies on permease led me to the conclusion that the exit mechanism although obeying passive kinetics is not a simple diffusion but rather a carrier mechanism. I also assumed that in permeaseless strains both, entry and exit go through this same carrier mechanism. The permeaseless cells provide the most convincing evidence: when equilibrated with a thiogalactoside until the internal concentration reaches the concentration in the medium, addition of glucose causes the thiogalactoside to leave the cell. This realizes a concentration difference and can be compared to an active extrusion of the thiogalactoside. In fact we interpret it as a backflow phenomenon, which, according to Dr. WILBRANDT, is the best evidence for carrier transport with a common carrier. It has to be stressed that a common carrier for glucose and galactoside does not mean a common permease. In permease positive cells glucose also displaces a part of the accumulated thiogalactoside, this appears as an acceleration of the exit rather than inhibition of entrance. This lends some support to the identity of mechanism of exit in permease positive and of passive transport in permease negative strains.

In this carrier hypothesis, the acceleration of transport can be due either to an increased amount of total carrier, or if unidirectional, to a gradient of the carrier, caused by a different process (glucose uptake in the experiment quoted above).

In the case of galactose permease, HORECKER describes an increase of the passive exit rate when cells are grown in the presence of galactose. This would be interpreted as an induced synthesis of carrier and leads to imagine the carrier as a protein and its induction as a specific phenomenon. These statements are rather intriguing in view of the relative lack of specificity of the carrier assumed above and in view of the physical characteristics one should postulate for a carrier.

If a protein involved in the exit process does exist either as a carrier or fulfilling any other role, one should also expect a mutation in which the exit mechanism is lost or impaired. Such a mutant has never been detected. This of course does not imply that such protein does not exit, but can be explained alternatively by assuming that its loss by mutation would be lethal.

BÜCHER: Vielleicht wäre dies der Ort, um doch noch einmal die Wirkung der Herzglykoside aufzugreifen. Kann man aus ihrem Angriff am aktiven Transport ihre spezifische Herzwirksamkeit erklären? Ich habe gesehen, daß Herr USSING uns heute einige Bilder unterschlagen hat, auf denen Strophantin auftauchte. Hier scheint es sich also um eine ganz allgemeine Wirkung gehandelt zu haben, denn das wird wahrscheinlich an der Froschhaut gewesen sein. Wir haben doch diese spezifische Wirkung am Herzen; wir haben so viele Fachleute unter uns, ich würde diese Frage, in der Diskussion ist sie schon angesprochen worden, ich würde sie doch noch ganz kurz aufwerfen und fragen, ob Herr WILBRANDT oder jemand noch etwas dazu sagen will und kann.

WILBRANDT: Ich glaube, daß wir heute noch keine definitive Antwort auf die Frage geben können, ob diese Hemmwirkungen auf Ionentransporte mit der therapeutischen Wirkung von Digitalis etwas zu tun haben. Vor allem können wir die Frage nicht beantworten, welches die Größe ist, auf deren Veränderung es ankommt. Die Hinweise dafür, daß die Transportwirkungen in Beziehung stehen zur therapeutischen Wirkung, sind recht zahlreich. Erstens einmal sind die Größenordnungen der wirksamen Konzentrationen nicht weit voneinander entfernt. Die Konzentration des Ouabains mit halbmaximaler Hemmwirkung auf den Natriumtransport in Erythrocyten liegt bei 10^{-7} g/ml. Wir haben mit Konzentrationen von 10^{-9} und 10^{-10} noch deutliche Wirkungen gesehen. Das ist etwa die Größenordnung, in der man in vitro unmittelbare Digitaliswirkungen auf den Herzmuskel beobachtet (z. B. in den viel zitierten Versuchen von H. GOLD u. M. CATELL [Arch. intern. Med. **65**, 263 (1940)] am Papillarmuskel). Wichtiger ist wohl, daß die strukturellen Beziehungen sehr ähnlich sind. Die Abstufungen zwischen verschiedenen Glykosiden, die Abstufungen zwischen Glykosiden und Geninen, die Abhängigkeit von der Anwesenheit des Rings, von der Doppelbindung im Ring, die Unterschiede zwischen α- und β-Modifikationen zeigen weitgehende Übereinstimmung zwischen den Wirkungen auf die Ionentransporte und den Wirkungen auf das Herz. Eine Hauptschwierigkeit ist die, die von Herrn KLAUS heute vorgebracht worden ist. Es gibt eine Anzahl von Versuchen zur Frage, ob am Herzmuskel ein Kaliumverlust durch wirksame Digitaliskonzentrationen auslösbar ist.

In mehreren Untersuchungen ergab sich ein klarer Verlust erst in toxischen Konzentrationen. Für die Annahme, daß die Kaliumkonzentration in der Herzmuskelfaser die maßgebende Größe ist, bedeutet das eine beträchtliche Schwierigkeit. Ich persönlich glaube, daß das Calcium noch zu wenig untersucht ist. Ich erinnere an die klassischen Beobachtungen von OTTO LOEWI aus dem Jahre 1918 über die Sensibilisierung des Herzens gegen Calciumwirkung durch Herzglykoside. Eine erste Arbeit in dieser Richtung, die vielleicht zu einer Deutung führen könnte, ist von HOLLAND und SEKUL im Amer. J. Physiol. **197**, 757 (1959) veröffentlicht worden. Die Verfasser kommen zu dem Schluß, daß Herzglykoside den Einwärtsflux des

Calciums in den Herzmuskel beschleunigen. Das würde innen zu einer Anreicherung von Calcium führen können. Vielleicht ist hier der Punkt, wo schließlich einmal die beiden Beobachtungsgruppen in Berührung kommen werden.

GREEFF (Düsseldorf): Ich bin wie Herr WILBRANDT der Meinung, daß auch die therapeutische Wirkung der Digitalisglykoside mit einer Beeinflussung des aktiven Transports verbunden sein könnte. In den von Herrn KLAUS erwähnten Versuchen ist zu berücksichtigen, daß der positiv inotrope Effekt am isolierten Herzvorhofpräparat nicht unbedingt mit dem therapeutischen Effekt der Digitalisglykoside zu vergleichen ist; Voraussetzung für einen therapeutischen Effekt der Glykoside ist eine Insuffizienz des Herzens. Auch könnte der Einfluß der Glykoside auf den Ionenaustausch isolierter Herzpräparate (Tyrodelösung, 28° C) verschieden sein bzw. in einem anderen Dosierungsbereich liegen als am Herzen in situ.

NETTER (Kiel): Ich möchte eine Frage zu der Lage des Calciums stellen. Sie haben davon gesprochen — ich habe es aber leider nicht verstanden, wie das ausgeführt wurde, und habe auch die Autoren leider übersehen —, daß das Strophantin die Calciumaufnahme fördert, also offenbar doch einen Influx besorgt? Ich habe mir bisher immer vorgestellt, auf Grund einer ganzen Reihe von Erfahrungen, die ich im einzelnen nicht aufführen kann, daß das Calcium weitgehend außen auf lipoiden Grenzschichten liegt. Das würde ja an sich zu dem Punkt passen, den Sie heute morgen selber vertreten haben, daß also die Strophantusglykoside an der Außenfläche angreifen. Ich möchte nun fragen, wie ist das technisch gemacht und erscheint Ihnen sicher, daß das Calcium nun doch in das Strophantin hineingeht?

WILBRANDT: Daß die Zellmembranen im allgemeinen für Calcium durchlässig sind, ist sicher nicht richtig. Besonders überzeugend sind die schönen Versuche von A. L. HODGKIN und R. D. KEYNES am Riesenaxon [Physiol. (Lond.) **138**, 253 (1957)]. Die Möglichkeit, das Axoplasma auszupressen und unmittelbar zu analysieren, verhindert hier, daß man durch Bindung von Calcium an die Oberfläche der Membran getäuscht wird. Aus diesen Versuchen weiß man mit Sicherheit, daß das Calcium bei der Erregung durch die Membran in das Innere der Zelle eintritt und in der Erholung herausgepumpt wird (übrigens gegen einen sehr steilen Gradienten). Dann hat A. SHANES [J. gen. Physiol. **42**, 803 (1959)] eingehende Beobachtungen am quergestreiften Muskel mitgeteilt. Am Herzmuskel hat R. NIEDERGERKE [Experientia (Basel) **15**, 128 (1959)] Versuche veröffentlicht. Nach allen diesen Resultaten kann kein Zweifel darüber bestehen, daß ein Austausch von Calcium durch die Membran stattfindet. Daß die intracelluläre Calciumkonzentration einen wesentlichen Einfluß auf das contractile System hat, ist andererseits wohl bekannt. Vielleicht äußert sich Herr WEBER oder Herr HASSELBACH noch zu dieser Frage?

DECKER (Hannover): Die Beeinflussung des Elektrolyt-Transportes durch Strophantin veranlaßt mich zu der Frage, ob an der Digitaliswirkung am Herzkranken nicht auch eine direkte Wirkung auf die Ödeme via Elektrolyttransport beteiligt ist, die indirekt dem Herzen zugute kommt.

KLAUS (Mainz): Im Hinblick auf unsere bereits heute vormittag mitgeteilten Befunde über die Wirkung der Herzglykoside auf den K-Flux sei noch einmal betont, daß keine Beeinflussung des K-Austausches bei therapeutischen Konzentrationen erfolgt, daß dagegen beim Auftreten von toxischen Erscheinungen der K-Efflux gesteigert wird und der K-Influx geringfügig reduziert wird. Es ist hierbei kein gradueller Übergang festzustellen. Das von uns gewählte Kriterium für den therapeutischen Effekt war die positiv inotrope Wirkung.

Es besteht wohl kein Zweifel, daß dies zumindest *eine* therapeutische Wirkung der Herzglykoside ist und zwar in bezug auf das Herz die wichtigste Am Ganztier erfolgen natürlich noch Beeinflussungen anderer Vorgänge, die aber für die von uns untersuchten Zusammenhänge unwesentlich erscheinen.

Die von HOLLAND und SEKUL mitgeteilte Ca-Influxsteigerung am Herzen unter Glykosideinfluß wurde nur unter toxischen Bedingungen gemessen, die Präparate zweigten alle eine Kontraktur. Wie weit deshalb eine gesteigerte Ca-Aufnahme bei der therapeutischen Wirkung eine Rolle spielt, muß deshalb noch genauer untersucht werden. Wir haben bisher nur den Ca-Nettogehalt unter Glykosideinwirkung bestimmt und dabei in therapeutischen Konzentrationen eine signifikante Erniedrigung gefunden. Eine Deutung dieser Befunde ist vorerst nicht möglich; hierzu müssen noch Messungen des Ca-Umsatzes durchgeführt werden.

HILZ (Hamburg): Im Zusammenhang mit der Strophantinwirkung am Herzen möchte ich daran erinnern, daß nach neueren Ergebnissen auch Aldosteron als starker Herzwirkstoff bekannt wurde. Aldosteron hat sicherlich Wirkungen auf Permeationsvorgänge, wie sich aus Untersuchungen an der Aorta ergeben hat. Mißt man unter in vitro-Bedingungen den Einbau von S-35-sulfat in die Polysaccharidfraktion, so kommt es in Anwesenheit geringer Glucocorticoidkonzentrationen (10^{-7}—10^{-6} m) nach einer kurzen Induktionsperiode zu einer beträchtlichen Steigerung (bis 270%) der Sulfopolysaccharidbildung. Höhere Konzentrationen ($> 10^{-5}$ m) hemmen diesen Vorgang. Aldosteron hingegen zeigt ab 10^{-8} m Konzentration eine Hemmung der Sulfopolysaccharidmarkierung, die mit steigender Steroiddosis zunimmt. Diese Hemmung scheint auf Grund einer Permeationshemmung zustande zu kommen. Mißt man nämlich nicht nur die Markierung der Polysaccharidfraktion, sondern die gesamte Radioaktivität in der Aorta, so ist in Anwesenheit von Aldosteron diese Gesamtaktivität sehr stark erniedrigt und die verminderte Sulfopolysaccharidmarkierung offensichtlich eine Folge der verminderten Sulfataufnahme. Ganz andere Verhältnisse liegen vor bei der Glucocorticoidwirkung. Hier wird der Gesamtgehalt des Gefäßes an S 35 gegenüber der Norm nicht verändert. Die Steigerung der Sulfopolysaccharidbildung ist sehr wahrscheinlich das Resultat einer gesteigerten Enzymaktivität.

Ob die Aldosteronwirkung die Permeabilität der bindegewebigen Grundsubstanz betrifft oder die Durchlässigkeit der Bindegewebszellen ist nicht klar.

GRAB (Gießen): Ich bin nicht ganz so zuversichtlich wie Herr Prof. WILBRANDT, ob die Wirkung von Strophantin auf das insuffiziente Herz verständlich werden kann auf Grund der Strophantineinwirkung auf die Membrandurchlässigkeit. Es wäre sehr befriedigend, wenn dies möglich wäre; aber einige Befunde scheinen mir nicht ganz zueinander zu passen. Strophantin wirkt auf die Zellmembranen auf der Außenseite, an der Herzmuskelfaser muß es in die Zelle eindringen, um sich mit der Muskelsubstanz zu verbinden. Strophantin wirkt auf die Membrandurchlässigkeit ziemlich rasch, auf das insuffiziente Herz erst nach längerer Latenzzeit. Für seine Wirkung auf die Membran braucht man wesentlich höhere Konzentrationen von Strophantin als für seine Wirkung auf die Herzmuskelfaser, besonders bei der „Erhaltungsdosis" der Herzglykoside. Ich halte es aber für durchaus denkbar, daß sich die Wirkung des Strophantins auf die Membranen und Muskelfasern miteinander verknüpfen lassen, aber wohl kaum ohne besondere Interpretation auf Grund neuer Untersuchungen.

Hasselbach (Heidelberg): Ich will nun dieses Kapitel abschließen, weil Wilbrandt auf uns angespielt hat. Ich will sagen, daß der Weg von der Membran zum contractilen Eiweiß weit ist und daß man eigentlich, wenn man die Veränderungen an der Membran kennt und Beziehungen hergestellt hat zwischen Permeabilität und Wirkung einer Substanz, noch nichts darüber aussagen kann, wie diese Substanz am contractilen Eiweiß wirkt; mit dem Stoffwechsel hat das sehr wahrscheinlich wenig zu tun, denn ob ein Muskel sehr hohe Leistungen vollbringt, muß nicht unbedingt von seinem Stoffwechsel abhängen. Ich habe dann noch eine persönliche Frage: Ich möchte Herrn Passow fragen, was er von dem Befund hält, den der Kollege aus München vorgetragen hat.

Es hatte doch den Anschein, daß es sich offenbar bei diesen Bestimmungen der Phosphatidsäure um Artefakte handeln kann.

Passow (Hamburg): Darf ich noch einmal fragen, um welches Argument es sich gehandelt hat?

Bader (München): Man kann Herzen perfundieren mit ^{32}P und bekommt Aktivitätsgipfel an der Stelle, an der die Phosphatidsäure ist. Genauso kann man aber die Phosphatide, den Rohextrakt in Chloroform-Methanol mit ^{32}P behandeln und bekommt den gleichen Aktivitätsgipfel; also ist es gar nicht nötig, den Umweg über die Perfusion zu machen. Eines scheint dabei eine Rolle zu spielen, das ist die Anwesenheit von Calcium; wahrscheinlich reagiert auch Magnesium oder Cadmium.

Passow: Darf ich fragen, mit welcher Methode Sie chromatographiert haben?

Bader: Mit Kieselsäuresäule. Hokin und Hokin arbeiteten papierchromatographisch.

Passow: Die Frage ist nun, ob das vergleichbare Bedingungen sind.

Bader: Die Extraktion von Hokin und Hokin ist nicht viel anders als unsere, und die Aufteilung, ob es nun an mit Kieselsäure imprägniertem Papier stattfindet oder an der Kieselsäuresäule, dürfte wohl kaum eine Rolle spielen; vorher ist das Ganze doch vorhanden, bevor es auf die Säule geht.

Holzer: Haben Sie die Identität dieser Gipfel weiter geprüft? Wenn zwei Substanzen in einem Chromatogramm an die gleiche Stelle laufen, so sagt das nicht viel.

Bader: Ja, dazu habe ich folgendes gemacht: Dieser Phosphatidsäuregipfel, der erste hier, ist immer an der gleichen Stelle bei unserer Methode. Die Methode ist so genau, daß er wirklich immer an ganz genau der gleichen Stelle kommt. Der Aktivitätsgipfel ist jedesmal, das ist bei über 50 Versuchen so gewesen, etwas vor diesem Gipfel, genauso die anderen Aktivitätsgipfel, die später kommen. Es decken sich nie genau diese Phosphorgipfel mit den Aktivitätsgipfeln.

Holzer: Wo findet sich das nachträglich zugesetzte Phosphat? Kann es nicht sein, daß Sie einfach Ihr zugesetztes Orthophosphat wieder im 1. Gipfel finden, sowohl nach der Passage als bei nachträglichem Zusatz? — Wo läuft Phosphat in Ihrem Chromatogramm hin, anorganisches Phosphat?

Bader: Es kann nur mitgenommen worden sein; wenn man anorganisches Phosphat in Chloroform-Methanol bringt und es dann auf die Säule gibt, kommt es nie heraus, nur wenn man mit Wasser eluiert, ganz zum Schluß.

Passow: Ich glaube, man kann zu den Versuchen von Hokin und Hokin etwas sagen: Sie haben den radioaktiven Austausch nur gemessen. Sie haben also radioaktives Phosphat hinzugetan und haben einmal in Gegenwart und einmal in Abwesenheit von Acetylcholin gemessen und fanden, wenn sie Acetylcholin drin hatten, mehr Radioaktivität in der Fraktion, die sie Phosphytidsäure nennen, und wenn sie kein Acetylcholin drin hatten fanden sie das nicht; sie haben aber in beiden Fällen dieselbe Menge an anorganischem Phosphat.

Bader: Wir haben folgendes gesehen: Durchströmt man Rattenherzen mit Insulin und Glucose oder ohne Insulin nur mit Glucose, so bekommt man bei kurzzeitigen Experimenten z. B. nach 10 min mit Insulin eine dreifache stärkere Markierung der Phosphatide als ohne Insulin; nach 30 min gleicht es sich wieder aus, nach 30 min ist kein Unterschied mehr zu sehen. Dies sind erst vorläufige Untersuchungen, die noch nicht veröffentlicht sind und die jetzt erst weitergeführt werden müssen.

Eine große Rolle scheint mir dabei jedoch eines zu spielen, das Calcium oder Magnesium, je nachdem, welches man nimmt. In Anwesenheit dieser divalenten Ionen geht das ^{32}P viel stärker an die Phosphatide als ohne diese. Gibt man z. B. Versin zu, so geht das Calcium vollständig aus den Phosphatiden heraus, das haben wir mit radioaktivem Calcium gemessen. Damit geht aber auch gleichzeitig das ^{32}P heraus, zwar nicht so gut, wie das Calcium, aber es geht heraus.

Bücher: Ich glaube, es bleibt uns nichts anderes übrig, als zur Kenntnis zu nehmen, daß der Herr Kollege die Versuche von Hokin und Hokin einer experimentellen Kritik unterzieht.

Bader: Die Frage ist, inwieweit wird bei der Extraktion das in jedem Fall vorhandene ^{32}P, das ja bei der Extraktion mit Chloroform-Methanol hereinkommt, eine Verfälschung der ganzen Befunde ergibt. Die nächste Frage ist die, wie man diese Verfälschung ausschalten kann.

BÜCHER: Darf ich da noch kurz eine Frage zu meiner Information stellen? Sind die Inositphosphatide völlig aus der Diskussion verschwunden? Mit diesen hat es doch an sich angefangen bei HOKIN und HOKIN.

BADER: Nein, HOKIN und HOKIN sind vom Inositphosphatid auf Phosphatidsäuren gekommen. Eine andere Frage ist, inwieweit es wirklich Phosphatidsäure ist und nicht Cardiolipin oder ähnliches. Am Herzen haben wir eigentlich nie freie Phosphatidsäuren gefunden.

BÜCHER: Und was ist mit den Inositphosphatiden?

BADER: Die Inositphosphatide sind bei uns wahrscheinlich in der gleichen Fraktion, wenn man mit Calcium behandelt. Hier kommt das Inositphosphatid, und dann bekommen wir hier genauso einen Aktivitätsgipfel wie bei den Phosphatidsäuren, der aufgeschoben ist. Dann haben wir also zwei.

DEBUCH (Köln): Ich möchte zu Ihrer Frage, wie man diese eventuelle Verfälschung ausschalten kann, folgendes sagen: Wenn Sie den rohen Chloroform-Methanol-Extrakt vom Herzen nehmen, den Sie chromatographisch trennen wollen, so möchte ich fast annehmen, daß es sich bei keinem Ihrer Phosphorgipfel um eine reine Substanz handelt. — Ich halte es doch für möglich, daß anorganisches Phosphat gelöst wird in der 1. Fraktion. Wir arbeiten eigentlich immer mit weitgehend vorgereinigten Fraktionen, und selbst dann finden wir manchmal mehrere Substanzen in einem Gipfel. Es ist also nicht so einfach, Ihre Befunde zu deuten.

BADER: Nun das ist bei beiden Methoden das gleiche, sowohl bei der Papierchromatographie als auch bei der Säule; man hat immer auf einem Fleck oder in einem Gipfel mehrere Substanzen drin. Man kann aber jetzt folgenden Trick machen: einmal die Säule ohne Calciumbehandlung oder mit vorheriger Versinbehandlung, dann bekommt man etwas andere Gipfel als mit Calciumbehandlung. Dadurch kann man ohne Calcium und mit Calcium die einzelnen Gipfel nochmals aufteilen. Dann teilen sich nun auch die Aktivitätsgipfel. Wir haben erst eine einzige Fraktion gefunden, die in der Cephalinfraktion ist, bei der tatsächlich der Aktivitätsgipfel und der Phosphorgipfel übereinstimmen über den ganzen Gipfelbereich.

The mechanism of active transport of ions in nerve and muscle fibres

By

R. D. Keynes

Agricultural Research Council Institute of Animal Physiology, Babraham, Cambridge

With 2 Figures

I ought to start by confessing that the subject I propose to discuss is a more restricted one than is suggested by my original title. "Mechanisms of ion transport" ought presumably to cover both the *downhill* ionic movements that occur when an impulse travels along a nerve or muscle fibre and the subsequent *uphill* movements which restore the concentration gradients. But there is good reason to believe (Hodgkin and Keynes, 1955) that the action potential or spike mechanism operates in parallel with, and can be dissociated from, the active transport or recovery mechanism. The immediate energy source for the spike mechanism appears to be the pre-existing ionic concentration gradient, so that the spike probably does not depend directly on cellular metabolism, but indirectly, by drawing its energy from a source which does require to be constantly replenished at the expense of metabolic substrates. As far as biochemistry is concerned, the obvious linkage is with the active transport mechanism, and on this occasion I therefore propose to say very little further about the spike mechanism.

Although it is fashionable nowadays to use the term "active transport" to describe all sorts of phenomena, it must be admitted that more often than not the main purpose of the phrase is to conceal our ignorance, at the molecular level, of the mechanisms involved. Even after surmounting the barrier of defining precisely what we mean by active transport, we are still almost wholly in the dark as to the kind of interactions that take place when molecules or ions cross cell membranes. Are we right to think in terms of carrier systems? I would remind you that no one has yet been able to isolate any chemical compound with a specific affinity for any actively transported substance which remotely approaches that of the living system. And although we are now able to distinguish clearly under the electron microscope a structure which we label as the membrane, we have yet to see anything which can plausibly be interpreted as a site of active transport. It may be salutary, before I embark on the main part of my talk, to remind you of one of the formidable difficulties that lie in the way of identifying chemically or making visible the transport mechanism. This arises from the expected paucity of active transport sites — they need in all probability only to be rather few and far between to account for the observed rates of transport, and are likely to occupy a correspondingly small fraction of the membrane area.

Possibly the best piece of evidence as to the density of ion transport sites comes from GLYNN's (1957) calculations for human erythrocytes, based on his observations of the minimum number of cardiac glycoside molecules necessary to block outward sodium transport. He arrived at a total of only 1000 sites for each erythrocyte. The total area of an erythrocyte is 120 μ^2, so that the area per site is 0.12 μ^2, and the average distance apart of the sites is about 0.35 μ. In an electron micrograph of a section 200 A.U. thick with sufficient magnification (e.g. $\times$400,000) to show the membrane as a pair of lines 2 mm apart (see ROBERTSON, 1960), there would only be one transport site in every 2,4 m of membrane, and it would always be hard to identify the sites with any confidence. In nerve and muscle fibres the spacing may well be of the same order, although my evidence for saying so is certainly not strong. HODGKIN and KEYNES (1957) found that the calcium entry during a single nerve impulse corresponded to 38 ions moving across 1μ^2 of membrane, so that if each site were activated by a single Ca^{2+} ion the area per site would be about 0.03 μ^2. This estimate refers, of course, to the sites responsible for downhill movement, but the active transport sites might be similarly spaced. However, it is clear that all these calculations might be wrong by at least a factor of 10 in either direction.

Before I discuss the few mechanisms for active transport of ions which have been proposed in any detail I should like to try to summarise the rather scanty experimental facts which any successful theory must take into account. The evidence, such as it is, comes mainly from work on nerve and muscle fibres, but it will be convenient to discuss other tissues at some points, since there are respects in which different types of cell seem to have similar properties.

a) Selectivity. There can be no doubt that nerve and muscle fibres are capable of actively transporting K^+ ions inwards and Na^+ ions outwards. Net transport in sodium-loaded frog muscle fibres allowed to recover in a K-rich medium was demonstrated clearly by DESMEDT (1953), and has since been studied extensively by Professor E. J. CONWAY and his colleagues (see CONWAY, KERNAN and ZADUNAISKY, 1961, etc.). Net transport in nerve fibres has not been investigated as extensively, but it is obvious that it must occur if only to make good the net gain of Na^+ and loss of K^+ which occur during prolonged tetanisation (KEYNES and LEWIS, 1951; for other references see TASAKI, TEORELL and SPYROPOULOS, 1961). The active transport mechanism has to be capable of moving Na^+ ions outwards from a situation where K^+ is the predominant cation, and K^+ ions inwards from a solution in which Na^+ predominates. In the squid giant axon the active sodium efflux, defined as the fraction of the efflux of labelled sodium which disappears on treatment with metabolic inhibitors, amounts to some 40 pmole/cm^2 sec; potassium efflux does not change perceptibly during inhibition, so that the efflux of K^+ moving outwards via the active channel is probably less than say 2 pmole/cm^2 sec. Since the internal [Na] is only about 1/8 of [K], the affinity of the transport system for Na on the inside of the membrane must be at least 160 $\times$ greater than its affinity for K. At the outside of the membrane the transport system displays a selectivity of the same order but in the other direction, K being preferred to Na.

An interesting point in this connexion is to enquire into the behaviour of the active transport system towards the other alkali metal ions. As far as lithium is concerned there is a marked contrast between the action potential mechanism (involving downhill ionic movements) which barely discriminates at all between Na^+ and Li^+ (HODGKIN and KATZ, 1949; KEYNES and SWAN, 1959b; HUXLEY and STÄMPFLI, 1951), and the active transport mechanism which, at least in frog muscle (KEYNES and SWAN, 1959b), moves Li^+ ions outwards only about 1/10 as fast as Na^+. There is suggestive evidence that both myelinated (CONNELLY, 1959) and non-myelinated nerve fibres (RITCHIE and STRAUB, 1957; GREENGARD and

STRAUB, 1958) may behave in a similar way. Other tissues where Li^+ seems to be transported more slowly than Na^+ are erythrocytes (MAIZELS, 1954), and frog skin (ZERAHN, 1955).

There is very little evidence about the relative affinity of the active inward transport mechanism as between K^+ and, say, Rb^+ ions. Using the effect on efflux of labelled Na from a frog muscle as an index, Dr. R. H. ADRIAN and I (unpublished) have obtained a suggestion that again the active transport system may display greater discrimination between K^+ and Rb^+ than the action potential mechanism (see KEYNES and ADRIAN, 1956). However, more work needs to be done on these lines.

b) Energy supplies. The two rival points of view on the energy source for active transport may be considered to be on the one hand CONWAY's redox pump theory, according to which the transport of ions is linked directly to transport of electrons by the cytochrome system, and on the other the more orthodox hypothesis that the linkage to metabolism is indirect, via energy-rich $\sim$P bonds generated by oxidative phosphorylation. Now that CONWAY et al. (1961) have agreed with FRAZIER and KEYNES (1959) that net sodium transport from frog muscle can take place in the presence of 2 mM-cyanide, and without any accompanying consumption of oxygen, it seems to me that the redox theory has to be abandoned in its purest form, though I agree with Professor CONWAY that it is still possible to envisage a system in which ATP provides an alternative source of energy to drive electrons, and coupled to them ions, along an oxidation-reduction chain. However, this lacks the attractive simplicity of the original redox theory, and does not seem to me to be obviously preferable to the other ways in which phosphate-bond energy might be used.

In squid axons, direct evidence for the intervention of $\sim$P bonds in activating uphill ion transport has been obtained by CALDWELL, HODGKIN, KEYNES and SHAW (1960a, b). By injecting various $\sim$P compounds into squid axons whose indigenous ATP and phosphagen had been depleted by treatment with cyanide or dinitrophenol we showed that active transport could be fully restored only by fairly large amounts of arginine phosphate or phosphoenolpyruvate. Injection of ATP restored a state in which the efflux of labelled Na appeared to be normal in the presence of 10 mM-K in the external solution, but in which its usual sensitivity to removal of external K was not exhibited and in which there was no active influx of K. This intermediate state can also be arrived at by partial inhibition of metabolism by DNP in an alkaline medium. It probably corresponds to the condition when internal [ADP] first begins to rise much above normal, which occurs as soon as supplies of phosphagen are exhausted. There is some evidence that frog muscle may behave in a similar way (see KEYNES, 1961).

Other evidence that active transport in nerve is driven by $\sim$P bonds has been obtained by GREENGARD and STRAUB (1959). In mammalian erythrocytes, which do not display oxidative metabolism and derive their energy entirely from glycolysis, there is also evidence of a similar kind (WHITTAM, 1958).

c) Blocking agents. The most widely used blocking agents are simply substances which interfere with metabolism (e.g. cyanide, azide, dinitrophenol, iodoacetate, oxygen lack), and which probably produce their effect by interfering with synthesis of ATP. Another effective way of blocking active transport is to lower the temperature. Both in nerve and in muscle the uphill ionic fluxes have a large temperature coefficient (HODGKIN and KEYNES, 1955; KEYNES and SWAN, 1959a), and near 0° C active transport is brought virtually to a standstill. Stimulated by a report (SVENSMARK, 1960) that ATPase is sensitive to substitution of deuterium for hydrogen in the water, I have recently examined the effect on the efflux of ^{22}Na from a frog muscle of exposing it to Ringer's solution made up in 99% pure D_2O. The efflux was reversibly reduced by about 40%; a curious rebound on restoring normal Ringer was observed in both the experiments that I did, so that I believe it to be genuine, but have no explanation for its occurrence.

In 1953 SCHATZMANN showed that cardiac glycosides like digoxin and ouabain blocked net ion movements in human erythrocytes, and since then these substances have been shown to have a similar action in a wide variety of other tissues. Thus, EDWARDS and HARRIS (1957) showed that ouabain in concentration 10^{-6} greatly reduced the Na efflux from frog muscle, while CALDWELL and KEYNES (1959) observed a similar action on squid giant axons. Our most interesting finding was that although the ouabain had a rapid and irreversible action when applied outside the axon, it had no effect when injected inside the axon, even in much larger quantities. I dare not conclude yet that there is a genuine asymmetry in the glycoside effect, because it is possible that the injected material is adsorbed on to some component of the axoplasm and never reaches the cell membrane. If the perfused axons recently described by BAKER, HODGKIN and SHAW (1961) turn out to pump sodium satisfactorily, this preparation will be ideal for settling the question.

d) Enzyme systems involved. In addition to the enzymes involved indirectly in metabolism, there appears to be an Mg-activated ATPase system which is very probably closely concerned in active transport. This has been studied by SKOU (1961) in preparations of sub-microscopic particles from crab nerve, and by POST (1961) and DUNHAM and GLYNN (1961) in erythrocyte ghosts. Perhaps the strongest reason for thinking that these systems form part of active transport mechanisms is that they are specifically inhibited by cardiac glycosides at the same concentration levels as the intact tissues. Moreover DUNHAM and GLYNN have found evidence for apparent competition between the glycosides and K^+ ions similar to that observed in whole erythrocytes (GLYNN, 1957a). The behaviour of the ATPase towards other alkali metal ions like Li^+ and Rb^+ has not yet been reported in detail, but is obviously worth investigating.

A major difficulty in examining the ATPase in preparations other than intact cells is that of exposing the system to a realistic ionic environment. The normal mode of working must be closely bound up with the fact that it is highly organised spatially, and is exposed to quite different ionic concentrations on its two sides. Findings from experiments in which the whole system is in the same ionic medium will need to be interpreted with caution.

e) Regulation. It does not seem necessary to suppose that the active transport mechanism is regulated in a complicated way. If the absolute size of the sodium efflux from a cell is directly proportional to the internal sodium concentration, then the internal [Na] will automatically be controlled at such a level that the sodium efflux exactly balances the sodium influx. This proposition raises two questions: is it justifiable to assume such a direct proportionality, and if so, what factors influence the value of the proportionality factor? Evidence supporting the view that sodium efflux varies linearly with internal [Na] was obtained for squid axons by HODGKIN and KEYNES (1956), but in frog muscle KEYNES and SWAN (1959a) found a less simple state of affairs. In muscles exposed to Li Ringer, the sodium efflux appeared to be proportional to the third not the first power of $[Na]_i$. I have since done a few more experiments on these lines and have found that at very low internal [Na]'s the relationship ceases to follow a cube law, and becomes nearly linear. I have also obtained some confirmation that the efflux changes observed in these experiments are related to $[Na]_i$ by imposing sudden changes in internal [Na] through alterations in external osmotic pressure with added dextrose. When $[Na]_i$ was increased $1.86\times$ by dissolving an extra 200 mM-dextrose in the Li Ringer outside, the Na efflux increased $3.6\times$, this rise being very close to that predicted from the curve relating efflux to $[Na]_i$.

These measurements on frog muscle were, of course, made under somewhat unphysiological conditions, since Li was substituted for Na in the external medium, and the internal [Na] was driven down to abnormally low levels. It seems probable that in practice the efflux/concentration relationship is *S*-shaped, and that at very high $[Na]_i$'s there is a tendency for the efflux to saturate (see FRAZIER and KEYNES, 1959). With such a relationship the efflux might appear to be directly proportional to $[Na]_i$ over a limited range in the region of its normal value.

The most obvious factor likely to vary the sodium influx *in vivo* is the electrical activity of the tissue, which will depend on outside events. *In vitro*, other factors come into play, mainly related to the failure of artificial media used in physiological experiments to imitate fully the normal body fluids. It may be worth pointing out that in experiments on net sodium movements it is always necessary to distinguish between effects on influxes and effects on effluxes: a rise in $[Na]_i$ may be caused either by a fall in Na efflux or by a rise in Na influx, involving two quite different mechanisms. Thus the rise in $[Na]_i$ in frog muscle treated with cyanide and iosoacetate seems to result more from a large increase in influx than from an inhibitory action on efflux (FRAZIER and KEYNES, 1959). And Miss M. MULLANEY and I have unpublished evidence which suggests that the 120/104 effects discussed by CONWAY et al. (1961) arise at least partly from effects of external osmotic pressure on sodium influx.

The most important factor which immediately changes Na efflux under experimental conditions is external [K]. All workers are agreed that in many tissues Na efflux rises with external [K], though I do not know of any quantitative study of the precise form of the relationship in nerve or muscle. This action is presumably related to the nature of the coupling between sodium and potassium movements, discussed in the next section. There is even less quantitative information about the effect of external sodium concentration on Na efflux. In frog muscles the efflux falls on substituting Li for Na (Keynes and Swan, 1959a), this being taken as evidence for the occurrence of an appreciable amount of "exchange diffusion" of Na through the membrane. However, in squid axons the reverse effect is normally observed, though under certain circumstances there may be some exchange diffusion (Caldwell et al. 1960b). This observation is undeniably rather hard to interpret.

Evidence as to hormonal effects on sodium efflux or potassium influx is scanty. I have tried in vain to alter the sodium efflux from frog muscle by applying aldosterone. However, Flückiger and Verzar (1954a, b) found some effects of cortical hormones and insulin on the ion permeability of rat diaphragm muscle, and there seems to be some unexplained interaction between glycogen and K content. Recently, Zierler (1959) has reported that insulin raises the membrane potential as well as the K content of rat skeletal muscle, but it is not yet entirely clear what is cause and what is effect.

f) Coupling between Na and K movements. Possibly the key question to which we seek an answer concerns the nature of the coupling between Na efflux and K influx. In principle, one can envisage two ways in which K^+ ions might be taken into the cell:

1. The sodium pump might be "electronic" and might create a potential across the cell membrane large enough to cause a net influx of potassium. The coupling between Na and K fluxes would in this case be purely electrical.

2. The pump might involve formation of covalent bonds between ions and carriers, so that Na and K cross the membrane as neutral complexes moving in equal amounts in opposite directions.

One should realise that these two possibilities are probably not incompatible with one another, and that the mechanism may actually involve a combination of both of them. Also, if one likes to propose a sufficient number of sub-compartments within the membrane, one can devise systems which are really electrically coupled, but in which operation of the pump does not appear to generate a net potential across the membrane.

There is unsatisfactorily little experimental evidence to enable one to choose between these two types of mechanism. On the whole, I should bet on a mainly neutral pump, despite the evidence obtained by HODGKIN and KEYNES (1956) on a squid axon that an injection of Na which raised the Na efflux also caused a slight rise in resting potential. Against this observation is the fact that when the Na efflux is blocked with metabolic inhibitors there is no obvious change in resting potential. It would be helpful to make accurate comparisons of resting potential and NERNST (equilibrium) potential for K during a net uptake of potassium by sodium-loaded tissues. For mechanism (1) a net uptake could only occur when the membrane potential was greater than E_K; for mechanism (2) the resting potential could remain smaller than E_K. In erythrocytes, this type of comparison argues strongly in favour of mechanism (2), since a net uptake of K occurs at a time when E_K is much larger than the membrane potential, which is generally assumed to be equal to E_{Cl}. Put in different terms, it is clear that in erythrocytes the value of the ratio $[K]_i/[K]_0$ can be much larger than that of $[Cl]_0/[Cl]_i$. There are indications that *Sepia* axons can absorb K when their resting potentials are less than E_K (HODGKIN and KEYNES, 1955), and I am hoping shortly to examine the behaviour of frog muscle in this respect.

I have already quoted the work of CONNELLY (1959) and GREENGARD and STRAUB (1958) on after-potentials in myelinated and non-myelinated nerve fibres. Their results are certainly relevant to this question, since it seems probable that the after-potentials arise from a faster extrusion of sodium following a burst of electrical activity. However, CONNELLY (1959) has interpreted his results as favouring the idea of an electrogenic pump, whereas GREENGARD and STRAUB (1958) interpreted theirs in terms of a neutral pump, the hyperpolarization being supposed to arise because K is absorbed fast enough to lower its external concentration slightly.

It is important to note that any coupling that may exist between Na and K movements is almost certainly not exactly 1/1.

Thus the absolute size of the active efflux of Na from a squid giant axon is, as I have already mentioned, around 40 pmole/cm^2 sec, whereas the size of the active influx of K is only about 15 pmole/cm^2 sec. The coupling ratio in this tissue is therefore 2 or 3/1. It would be unwise to assume that the ratio is fixed, and I suspect that any proposed mechanism ought really to provide for a variable coupling ratio, with some inward movement of Na balancing part of the Na efflux, i.e. with some exchange diffusion of Na. It is also conceivable that there are some anion movements coupled to the sodium extrusion.

g) Quantitative correlation between sodium efflux and energy consumption. In frog muscle CONWAY et al. (1961) find that the net extrusion of 4 Na^+ ions requires the consumption of 1 O_2 molecule; KEYNES and MAISEL (1954) arrived at the same figure for the efflux of labelled Na. If the ratio of $\sim$P bonds formed to O_2 molecules consumed is 6, this means that the ratio Na/$\sim$P is 0.67. It may be significant that our injection experiments with squid axons also gave a ratio close to 0.7 for injection of ATP, arginine phosphate and PEP. Remembering that active transport is probably not the sole energy-consuming process in operation during these experiments, these figures are not inconsistent with a stoichometrical 1/1 relationship between splitting of $\sim$P bonds and Na extrusion. The work done in extruding Na from nerve or muscle fibres is in the region of 3000 cal/mole, so that a 1/1 ratio does not correspond to an unreasonable value for the efficiency of the process.

In other tissues it is clear that the Na/$\sim$P ratio may be much larger. The figure for frog skin quoted yesterday by Professor USSING was Na/O_2 = 18, which corresponds to Na/$\sim$P = 3, and in kidney the ratio is certainly higher still. If the mechanism for sodium extrusion is similar in all cases, then one must assume that it is capable of adjustment so that fewer $\sim$P bonds are split when the energy barrier is low, as it is in kidney and as it may be in frog skin.

Proposed mechanisms for active ion transport

I have not much time left to discuss the mechanisms that have been proposed for active transport of sodium and potassium, but as I have little faith in any of the schemes suggested so far, this does not greatly matter. I ought first to mention the possibility that pinocytosis is involved. In certain special situations (e.g. ion transport in the gall bladder) this idea has its attractions, but as far as nerve and muscle are concerned pinocytosis seems to fall far short of explaining the experimental findings that I have just summarised. In particular, I do not see how pinocytosis could

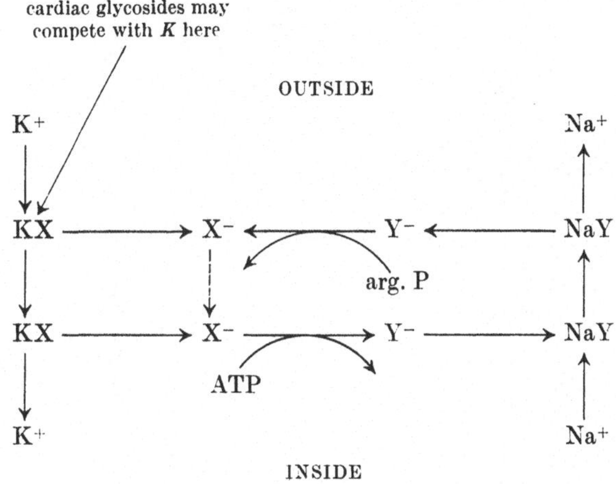

Fig. 1. Modified carrier scheme for coupled ion transport in nerve and muscle

achieve the necessary selectivity, nor how it could account for the coupling between inward and outward ionic movements.

Almost all the other mechanisms that have been considered have been carrier schemes of one sort or another. Many of them have already been described admirably by Professor Netter, so that I do not need to present a comprehensive list. I wish simply to elaborate slightly the carrier system discussed by Glynn (1957b), and to produce a new scheme due mainly to R. E. Davies (see Davies and Keynes, 1961). The point about the modified version of Glynn's system shown in Fig. 1 is that it would explain the behaviour of the active transport mechanism during partial inhibition of metabolism, when ATP is available but arginine phosphate is not. Thus if energy from arginine phosphate was needed directly at the outer side of the system to produce the K carrier, provision of ATP alone would result in a piling up of the Na carrier and in the occurrence of a sodium-coupled Na efflux as is observed. However, this idea has some obvious drawbacks, and it could equally well be supposed that there are two cyclical carrier systems operating in parallel — a sodium-coupled system active when the [ATP]/[ADP] ratio has an intermediate value, and a potassium-coupled system which only comes into action when the [ATP]/[ADP] ratio is high. I have also indicated in Fig. 1 the possibility that the carrier X might be able to return through

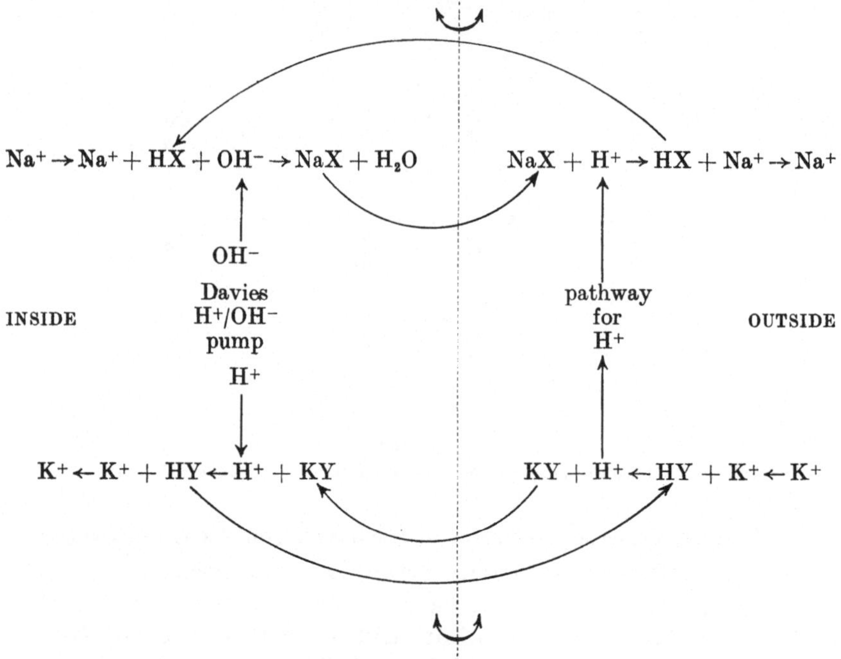

Fig. 2. Extrusion of Na^+ and absorption of K^+ coupled to R. E. DAVIES's H^+/OH^- pump

the membrane as X^-, not directly combined with K^+; this feature converts the system from a tightly coupled and electrically neutral pump to an electrogenic sodium pump. Both pathways for the return of X may, of course, operate at the same time.

Finally, Fig. 2 shows a way in which transport of Na and K might be linked to a system which uses phosphate-bond energy to bring about a continual separation of H^+ and OH^- ions (DAVIES, 1957). The carriers X and Y are supposed to form part of a structure which can rotate within the cell membrane. If HY had a pK of about 4 and HX a pK of about 6, then a flow of H^+ ions down the p_H gradient created by the H^+/OH^- pump would result in an inward transport of K^+ and an extrusion of Na^+. This idea has been described in detail elsewhere (Davies and Keynes, 1961).

References

Baker, P. F., A. L. Hodgkin and T. I. Shaw: Perfusion of the giant nerve fibres of Loligo. J. Physiol. (Lond.) (in the press) 1961.

Caldwell, P. C., A. L. Hodgkin, R. D. Keynes and T. I. Shaw: The effects of injecting 'energy-rich' phosphate compounds on the active transport of ions in the giant axons of Loligo. J. Physiol. (Lond.) **152**, 561—590 (1960a).

Caldwell, P. C., A. L. Hodgkin, R. D. Keynes and T. I. Shaw: Partial inhibition of the active transport of actions in the giant axons of Loligo. J. Physiol. (Lond.) **152**, 591—600 (1960b).

Caldwell, P. C., and R. D. Keynes: The effect of ouabain on the efflux of sodium from a squid giant axon. J. Physiol. **148**, 8—9 P (1959).

Connelly, C. M.: Recovery processes and metabolism of nerve. Rev. mod. Phys. **31**, 475—484 (1959).

Conway, E. J., R. P. Kernan and J. A. Zadunaisky: The sodium pump in skeletal muscle in relation to energy barriers. J. Physiol. (Lond.) **155**, 263—279 (1961).

Davies, R. E.: Gastric hydrochloric acid production -- the present position. In: Metabolic Aspects of Transport across Cell Membranes, ed. Q. R. Murphy. University of Wisconsin Press 1957.

Davies, R. E., and R. D. Keynes: A coupled sodium-potassium pump. In: Symposium on Membrane Transport and Metabolism, ed. A. Kleinzeller and A. Kotyk. Czechoslovak Academy of Sciences and Academic Press 1961.

Desmedt, J. E.: Electrical activity and intracellular sodium concentration in frog muscle. J. Physiol. (Lond.) **121**, 191—205 (1953).

Dunham, E. T., and I. M. Glynn: Adenosinetriphosphatase activity and the active movements of alkali metal ions. J. Physiol. (Lond.) **156**, 274—293 (1961).

Edwards, C., and E. J. Harris: Factors influencing the sodium movement in frog muscle with a discussion of the mechanism of sodium movement. J. Physiol. (Lond.) **135**, 567—580 (1957).

Flückiger, v. E., u. F. Verzar: Der Einfluß des Kohlenhydratstoffwechsels auf den Natrium- und Kaliumaustausch des überlebenden Muskels. Helv. physiol. pharmacol. Acta, **12**, 50—56 (1954).

Flückiger, v. E., u. F. Verzar: Die Wirkung von Corticosteroiden auf den ^{24}Natrium- und ^{42}Kalium-Austausch des Muskels normaler Tiere. Helv. physiol. pharmacol. Acta **12**, 57—62 (1954).

Frazier, H. S., and R. D. Keynes: The effect of metabolic inhibitors on the sodium fluxes in sodium-loaded frog sartorius muscle. J. Physiol. (Lond.) **148**, 362—378 (1959).

GLYNN, I. M.: The action of cardiac glycosides on sodium and potassium movements in human red cells. J. Physiol. (Lond.) **136**, 148—173 (1957a).

GLYNN, I. M.: The ionic permeability of the red cell membrane. Progr. Biophys. 8, 241—307 (1957b).

GREENGARD, P., and R. W. STRAUB: After-potentials in mammalian non-myelinated nerve fibres. J. Physiol. (Lond.) **144**, 442—462 (1958).

GREENGARD, P., and R. W. STRAUB: Effect of frequency of electrical stimulation on the concentration of intermediary metabolites in mammalian non-myelinated fibres. J. Physiol. (Lond.) **148**, 353—361 (1959).

HODGKIN, A. L., and B. KATZ: The effect of sodium ions on the electrical activity of the giant axon of the squid. J. Physiol. (Lond.) **108**, 37—77 (1949).

HODGKIN, A. L., and R. D. KEYNES: Active transport of actions in giant axons from Sepia and Loligo. J. Physiol. (Lond.) **128**, 28—60 (1955).

HODGKIN, A. L., and R. D. KEYNES: Experiments on the injection of substances into squid giant axons by means of a microsyringe. J. Physiol. (Lond.) **131**, 592—616 (1956).

HODGKIN, A. L., and R. D. KEYNES: Movements of labelled calcium in squid giant axons. J. Physiol. (Lond.) **138**, 253—281 (1957).

HUXLEY, A. F., and R. STÄMPFLI: The effect of potassium and sodium on resting and action potentials of single myelinated nerve fibres. J. Physiol. (Lond.) **112**, 496—508 (1951).

KEYNES, R. D.: Le transport actif des ions dans le muscle et le nerf. Actualités Neurophysiologiques, 3e Série, ed. A. M. MONNIER. Paris: Masson 1961.

KEYNES, R. D., and R. H. ADRIAN: The ionic selectivity of nerve and muscle membranes. Faraday Soc. Disc. **21**, 265—271 (1956).

KEYNES, R. D., and P. R. LEWIS: The sodium and potassium content of Cephalopod nerve fibres. J. Physiol. (Lond.) **114**, 151—182 (1951).

KEYNES, R. D., and G. W. MAISEL: The energy requirement for sodium extrusion from a frog muscle. Proc. Roy. Soc. B, **142**, 383—392 (1954).

KEYNES, R. D., and R. C. SWAN: The effect of external sodium concentration on the sodium fluxes in frog skeletal muscle. J. Physiol. (Lond.) **147**, 591—625 (1959a).

KEYNES, R. D., and R. C. SWAN: The permeability of frog muscle fibres to lithium ions. J. Physiol. (Lond.) **147**, 626—638 (1959b).

MAIZELS, M.: Active cation transport in erythrocytes. Symp. Soc. exp. Biol. 8, 202—227 (1954).

POST, R. L.: Membrane adenosine triphosphatase as a part of a system for active sodium and potassium transport. In Symposium on Membrane Transport and Metabolism, ed. A. KLEINZELLER and A. KOTYK. Czechoslovak Academy of Sciences and Academic Press 1961.

RITCHIE, J. M., and R. W. STRAUB: The hyperpolarization which follows activity in mammalian non-medullated fibres. J. Physiol. (Lond.) **136**, 80—97 (1957).

ROBERTSON, J. D.: The molecular structure and contract relationships of cell membranes. Progr. Biophys. **10**, 343—418 (1960).

SCHATZMANN, H. J.: Herzglykoside als Hemmstoffe für den aktiven Kalium- und Natrium-Transport durch die Erythrocytenmembran. Helv. physiol. pharmacol. Acta **11**, 346—354 (1953).

SKOU, J. C.: The relationship of a (Mg^{2+} + Na^+)-activated, K^+-stimulated enzyme or enzyme system to the active linked transport of Na^+ and K^+ across the cell membrane. In Symposium on Membrane Transport and Metabolism, ed. A. KLEINZELLER and A. KOTYK. Czechoslovak Academy of Sciences and Academic Press 1961.

SVENSMARK, O.: The effect of deuterium oxide on the active state of muscle. Acta physiol. scand. 50, Suppl. **175**, 148—149 (1960).

TASAKI, I., T. TEORELL and C. S. SPYROPOULOS: Movement of radioactive tracers across squid axon membrane. Amer. J. Physiol. **200**, 11—22 (1961).

WHITTAM, R.: Potassium movements and ATP in human red cells. J. Physiol (Lond.) **140**, 479—497

ZERAHN, K.: Studies on the active transport of lithium in the isolated frog skin. Acta physiol. scand. **33**, 347—358 (1955).

ZIERLER, K. L.: Effect of insulin on membrane potential and potassium content of rat muscle. Amer. J. Physiol. **197**, 515—523 (1959).

Diskussion

Diskussionsleiter: HOLZER, *Freiburg*

Thank you very much Dr. KEYNES for this highly interesting paper. The discussion of your paper will be opened by a paper from Dr. BREUER. Darf ich Sie, Herr Dr. BREUER zu ihrer Diskussionsbemerkung bitten.

BREUER (Bonn): Ich möchte an die Ausführungen von Herrn Dr. KEYNES anschließen und kurz auf einige Zusammenhänge zwischen dem Kationentransport und dem Energiestoffwechsel in Parenchymgewebe eingehen. Es wurde bereits darauf hingewiesen, daß die meisten tierischen Gewebe und Zellen nur unter aeroben Bedingungen Ionengradienten aufrechterhalten können. Unsere Untersuchungen, die zum Teil mit Dr. WHITTAM in Sheffield und Bonn durchgeführt wurden, haben gezeigt, daß die Mucosa der Vesiculardrüsen (sog. Samenblasen) des Meerschweinchens sowohl unter aeroben als auch unter anaeroben Bedingungen Ionen aktiv transportieren können. Das besondere Verhalten von Vesiculardrüsenschnitten wird bei einem Vergleich mit Nierencortexschnitten des Meerschweinchens deutlich. Inkubiert man die Schnitte in vitro für 60 min in Krebs-Ringer-Phosphatlösung, so verlieren sowohl die Cortexschnitte als auch die Mucosaschnitte unter aeroben Bedingungen zunächst einen Teil des intracellulären Kaliums; nach diesen anfänglichen Veränderungen beginnen dann die Zellen, Kalium wieder zu akkumulieren, und während der folgenden 40–50 min besteht hinsichtlich des Kaliumgehaltes ein Gleichgewicht. Der Konzentrationsgradient für Kalium zwischen dem

Gewebe und dem extracellulären Raum beträgt bei den Nierencortexschnitten 14 und bei der Vesiculardrüsenmucosa 18. Unter anaeroben Bedingungen verlieren wiederum beide Gewebe zunächst einen Teil ihres Kaliumgehaltes; während jedoch die Vesiculardrüsen für Kalium einen Konzentrationsgradienten von 14 aufrechterhalten können, sind die Nierenschnitte nicht mehr in der Lage, den Kaliumverlust rückgängig zu machen. Nach 40 minütiger Inkubation ist der Kaliumgradient auf 5,5 abgesunken. Ein ähnliches Verhalten wurde für Natrium beobachtet. Unter aeroben Bedingungen nimmt zwar der Natriumgehalt während der Inkubation in beiden Geweben gegenüber den Ausgangswerten zu, doch werden immerhin Gradienten von 0,63 für die Nierencortex und von 0,47 für die Vesiculardrüsenmucosa beobachtet. Unter anaeroben Bedingungen beträgt der Natriumgradient zwischen der Vesiculardrüsenmucosa und der Salzlösung noch 0,53, während er bei den Nierencortexschnitten auf 0,78 abgesunken ist. Diese Versuche lassen die Sonderstellung der Mucosa der Vesiculardrüsen erkennen, wonach dieses Gewebe sowohl unter aeroben als auch unter anaeroben Bedingungen Ionen aktiv transportieren kann.

Um festzustellen, ob die Elektrolytveränderungen, die bei einer Reduktion der Inkubationstemperatur auftreten, mit entsprechenden Änderungen in der Energiezufuhr korreliert werden können, wurden Mucosaschnitte der Vesiculardrüsen bei verschiedenen Temperaturen unter aeroben und anaeroben Bedingungen inkubiert. Der Q_{O_2} beträgt etwa 7 bei 37° und nimmt mehr als 50% ab, wenn die Temperatur auf 27° reduziert wird. Die Ionenkonzentrationen ändern sich dagegen nur um etwa 7–14% im gleichen Temperaturbereich. Offenbar ist die bei 27° verfügbare Stoffwechselenergie ausreichend, um einen aktiven Ionentransport sicherzustellen. Ähnliche Ergebnisse werden unter anaeroben Bedingungen erhalten. Auch hier ist die Reduktion der anaeroben Glykolyse wesentlich deutlicher als die der Ionengradienten. Diese Befunde lassen erkennen, daß die Elektrolytkonzentrationen zwar von der Temperatur abhängig sind, nicht aber in einer einfachen Beziehung zur Größe der Atmung oder anaeroben Glykolyse stehen.

Die Tatsache, daß Konzentrationsgradienten für Kalium und Natrium sowohl unter aeroben als auch unter anaeroben Bedingungen aufrechterhalten werden, eliminiert die Möglichkeit, daß der Mechanismus des Kationentransportes von der Sauerstoffaufnahme oder dem Elektronentransport im Cytochromsystem abhängig ist; es lag die Vermutung nahe, daß energiereiche Phosphatverbindungen dem Kationentransport die notwendige Energie zuführen. Deshalb wurde das Verhalten der Elektrolyte und von ATP unter verschiedenen Bedingungen untersucht. Die Versuche ergaben, daß sowohl unter aeroben als auch unter anaeroben Bedingungen bei 37° nur geringfügige Änderungen der Ionenkonzentrationen und der ATP-Konzentration auftreten. Nach Zusatz von Stoffwechselinhibitoren (DNP, Jodacetat) und nach Inkubation bei 0° kam es zu vergleichbaren Verlusten von Kalium und ATP sowie zu einer entsprechenden Aufnahme von Natrium. Daraus darf geschlossen werden, daß ATP in Vesiculardrüsen sowohl unter aeroben als auch unter anaeroben Bedingungen als Energiequelle benutzt wird.

Von Interesse erschien die Frage, in welchem Umfange die intracelluläre Kaliumkonzentration in Vesiculardrüsenmucosa von der Natriumkonzentration im extracellulären Raum abhängig ist. Der intracelluläre Kaliumgehalt nimmt kaum ab, wenn das extracelluläre Natrium durch Cholinchlorid oder Rohrzucker ersetzt wird. Falls diese Kaliumretention durch ein Donnan-Gleichgewicht bedingt ist, sollte der Zusatz von Stoffwechselinhibitoren keine Wirkung haben. In der Tat führt der Zusatz von DNP und Jodacetat nur bei den Versuchen in Krebs-Ringerlösung, nicht aber bei denjenigen in Cholinchlorid- oder Rohrzuckerlösungen zu einer Abnahme des Kaliumgradienten. Möglicherweise bewirken nichtdiffusible, intracelluläre Anionen ein Donnan-Gleichgewicht, das für die hohen Kaliumkonzentrationen nach Inkubation in den natriumfreien Lösungen verantwortlich ist.

Abschließend sei kurz auf eine Beobachtung eingegangen, die bei der Untersuchung strahlenbedingter Elektrolytveränderungen gemacht wurde. Es zeigte sich, daß die Kaliumgradienten in Nierencortexschnitten mit zunehmender Strahlendosis (3 bis 12×10^5r) erheblich abnehmen. Gleichzeitig steigt die Sauerstoffaufnahme um 30–40% an. Ähnliche Wirkungen werden auch nach Zusatz von DNP beobachtet; demnach konnte man annehmen, die Strahlung übe eine entkoppelnde Wirkung aus. Diese Möglichkeit ließ sich jedoch weitgehend ausschließen, so daß sich hinsichtlich des Wirkungsmechanismus der angeführten Veränderungen folgende Erklärung anbietet. Durch die Bestrahlung wird die Zellpermeabilität verändert, wobei bestimmte Enzyme freigesetzt werden und ungehinderten Zugang zu solchen Stellen haben, die sie normalerweise nur langsam oder gar nicht erreichen; auf diese Weise könne eine Zunahme der substratabhängigen, endogenen Atmung erfolgen. Andererseits kann diese Permeabilitätsänderung gleichzeitig zu einer Beeinträchtigung oder zu einem Versagen jenes Transportmechanismus führen, der für die Aufrechterhaltung von Konzentrationsgradienten verantwortlich ist. Infolgedessen wird die zusätzlich gewonnene Atmungsenergie für den aktiven Ionentransport nicht mehr nutzbar gemacht.

Diese wenigen Beispiele sollten zeigen, daß auch Gewebeschnitte zur Untersuchung des aktiven Ionentransportes geeignet sind.

HOLZER: Ich danke Ihnen für Ihren Beitrag, Herr Dr. BREUER, und eröffne die Diskussion. Ich bitte die Fragen an Professor KEYNES in Englisch zu stellen, falls eine deutsche Frage an Herrn KEYNES gerichtet wird, so hat sich Professor WILBRANDT freundlicherweise bereit erklärt, diese ins Englische zu transformieren. Vielleicht könnte als erstes das Problem des Ersatzes von Li und der Transport von Lithium und Natrium diskutiert werden. Herr Professor NETTER war dazu angesprochen worden.

Have you, Dr. KEYNES, a coworker, who perhaps would tell us something of this problem?

KEYNES: Dr. STRAUB has done this work.

STRAUB (KÖLN): I should like to discuss some experiments on the question of the coupling between Na extrusion and K uptake. As was mentioned by Dr. KEYNES, it is not known for certain whether in nerve fibres active extrusion of sodium ions is always tightly coupled with uptake of potassium ions, as it is in erythrocytes, or whether sodium can be extruded without simultanous uptake of potassium. The long-lasting rise in membrane potential which follows a period of repetitive activity in small nerve fibres gives a means for studying this question. Experiments of Dr. CONNELLY and of Dr. RITCHIE and myself have shown that this post-tetanic hyperpolarization is caused by active extrusion of sodium ions. Qualitatively, either the coupled or the uncoupled Na extrusion mechanism could produce the hyperpolarization: if Na extrusion is coupled with K uptake, increased extrusion of Na after activity would lead to a lowering of the K concentration outside the axon membrane and would thereby raise the membrane potential; alternatively, uncoupled Na extrusion could directly produce the post-tetanic hyperpolarization. In non-myelinated fibres the post-tetanic hyperpolarization could be explained by the coupled pump, qualitatively at least. In myelinated fibres, however, Dr. CONNELLY found it necessary to postulate an electrogenic pump for explaining the slow build-up of hyperpolarization which he observed during low frequency stimulation. His explanation was based on the assumption that sodium extrusion took place at the nodes of Ranvier only, and that the myelin-covered portions of the axons did not play any role in the maintenance of the membrane potential. So far there is no direct evidence for this assumption, but rather some observations on the effect of TEA on the membrane potential suggest that the myelin-covered parts may contribute to the membrane potential. Thus, the hyper-polarization described by Dr. CONNELLY could possibly be explained by a coupled pump removing K from the space between axon and myelin sheath.

More evidence for uncoupled Na extrusion came from studies which I made recently in myelinated fibres. They showed that the amplitude of the post-tetanic hyperpolarization was directly proportional to the membrane resistance, over a wide range of membrane resistance. Further, Dr. BÖHM and I observed that the amplitude of the hyperpolarization was not affected by removal of K from the extracellular solution. Moreover, post-tetanic hyperpolarizations of more than 10 mV could be recorded after very short periods of activity (30 sec at 50 c/s) during which little sodium could have diffused into the internode. Calculation showed that if the membrane potential and the intracellular ion concentrations found in the resting nerve were maintained by a coupled pump, the post-tetanic rise in the intracellular Na concentration in the myelincovered portions would have been too small to account for the size of the recorded post-tetanic hyperpolarization. A similar calculation showed that an uncoupled Na extrusion mechanism operating at the nodes could probably produce a hyperpolarization of the observed size.

HOLZER: Is there any comment on this?

KEYNES: This is very nice evidence.

BÜCHER (Marburg): Most discussion on energy transformation is done on the basis of normal free energy or normal potentials and it seems to me an important feature of your outstanding lecture that you showed a correlation between the steady state free energy of the ATP system and energy requiring function of the transport of the ions. What interests me is, and I am sure you have some reasons for it, that you discussed only the proportion of ADP to ATP and omitted the proportion of the system to inorganic phosphate.

Of these three parameters you discuss only two. The effect of arginine phosphate would not mean that the steady state concentration of inorganic phosphate is not important in this respect to the free energy of the ATP system.

KEYNES: I think that is a very good point, and I don't think I do have any reason for having ignored the inorganic phosphate concentration. It seems to me that in a way the most interesting question that these results have raised, is whether the energy for active transport could perhaps be fed directly from the phosphagen into the system or whether it has all to come via ATP. What I really would like biochemists to give us as a tool is a specific inhibitor for the enzyme arginine phosphoryl transferase; because we have done recently experiments in which we took axons poisoned with cyanide and injected arginine phosphate, and we found that within a minute or two, as soon we could measure it, the ATP concentration was at its normal level. What we want to be able to do is to distinguish between mechanisms which depend strictly on variations of the ATP/ADP ratio and a possible direct feeding of energy from arginine phosphate straight into the system. But since the arginine phosphate and this ratio appear to be virtually in instantaneous equilibrium we find ourselves quite unable to distinguish between the two possibilities. If anyone could suggest a way in which we could block the arginine phosphate and not allow it to rephosphorylate the ADP in the system, that would be very nice.

BÜCHER: Some indirect evidence should come out from a study of the three parameters ADP, inorganic phosphate, free arginine and arginine phosphate in relation to biological activity.

KEYNES: We have done that to a certain extent, but I quite agree that it is something we ought to do more. We have tried the effects of injecting arginine rather than arginine phosphate, but we have not tried to inject inorganic phosphate.

WILBRANDT (Bern): I was greatly interested in the modified scheme of Shaw's, but there is one point that was not quite clear.

The action of ATP is inside as you have shown with your injection experiments, and by the addition of ATP to the external surface where it did not act, but your experiments with arginine phosphate were injection experiments too. There I see a difficulty.

KEYNES: This is why I said I was not sure that I believed in carriers, because to make this mechanism work you have got to suppose that the arginine phosphate can pass from inside to the place where it acts, but cannot get at it from the outside. We did controls in which the arginine phosphate was applied outside, and it had no action at all. We did not, as a matter of fact, ever do the experiment of providing ATP inside and simultanously providing arginine phosphate outside. This would possibly restore K-coupled transport, but I would be very surprised of it did. But you can have quite a different alternative to this scheme. It would be that you have got two forms of cycle in parallel. One of them operates when the ADP/ATP ratio is very small to give you a coupled sodium-potassium movement; when this ratio has an intermediate value then for some reason it gives only a sodium-sodium exchange, and when the ratio becomes very large, then it stops moving sodium outwards at all. I don't think you can discriminate at present between these hypotheses. But as I said, I am not really convinced by any of them.

JATZKEWITZ (München): Dr. KEYNES, I would like to ask you, what do you think about the participation of phosphoproteins in active transport of nerve as postulated by Heald?

KEYNES: Well, since we have no idea at all what the carriers are, I think it is quite a nice idea that they could be phosphoproteins or something of that sort. The difficulty is, as I said at the beginning, to find any substance which can discriminate between sodium and potassium to the extent that we observe, whether or not it is phosphorylated. I think that the serious drawback to the rather nice scheme that Dr. HOKIN produced is that his cycle is not a K-coupled system. I think also that he admits that in point of fact it has very little specific affinity for sodium as against potassium, so that you need something else, and I am perfectly prepared to accept phosphoproteins as a possibility if they can be shown to have the necessary high specificity.

Mandel (Strasbourg): I should like to mention, that the fraction that Heald refers to the phosphoproteid fraction, is really a very complex fraction in which we found phosphate of phosphoproteins, phosphate of phosphatidylpeptides and phosphate of inositolphosphates. Now it is possible that the inositolphosphate of the phosphatidyl peptides are connected with the phenomenon more closely than the phosphoproteins.

Keynes: This is a very attractive idea. I hope that people will look for other phosphorylated compounds which could act like ATP. It always seems to me that we have to consider another little something beyond ATP, and I am sure the study of these phosphate compounds will in the end prove extremely important.

Holzer: Wir müssen etwas konzentriert diskutieren, weil wir mit der Zeit knapp sind.

Bramstedt (Hamburg): I think the experiments you showed us here are very interesting from the standpoint of comparative biochemistry.

We know that there exist animals which have energy rich phosphate as arginine phosphate and as ATP, and now it is interesting that there are some species of animals such as the echinoderm and tunicates which have both arginine phosphate and ATP. Now as you have shown, the rise to the original level can occur after poisoning with cyanide and removing the cyanide, but you did not find a complete rise to the original level if you injected ATP; and in the other experiment you showed if you used arginine phosphate, you came to the same level. Now I would like to ask, have there been any experiments with those animals containing both energy rich compounds because it would be of interest if you could find also the rise to the same original level if you use ATP or if you use arginine phosphate, or are such experiments not done?

Keynes: I think the quick answer to that is that if they have not got lage axons it will be difficult to do the experiments. But it is a nice point.

Holzer: I should like to make only one remark on the problem of the ATP-ase. About two years agothere was a symposium on regulation of metabolism, and Dr. Lippman, Dr. Krebs, Dr. Lynen and Dr. Martius and others discussed in about half an hour the problem whether there exists a single enzyme which makes ATP, given ADP and phosphate, or AMP and diphosphate, and they agreed all together, both these people and a lot of others, that there does not exist a single ATP-ase, that ATP-ase is only a sum of reactions which have the above mentioned „Bilanz": ATP, ADP and phosphate.

Keynes: Yes, I quite agree.

Wilbrandt: I have only a short question. I think I remember correctly that from the dependence of the spike potential on the membrane potential.

HODGKIN calculated that the carrier should have three negative charges or something like that.

KEYNES: Yes.

WILBRANDT: Now do you see a connection between your finding that the active transport of sodium grows with the third power and this postulated charge of the carrier?

KEYNES: Well, the number comes out of the same order but I think that could be a mere coincidence. I think that the carrier which serves the action potential is quite different from the carrier which works the pump. And it does seem to me that probably both of them are multiply charged; but there are other possibilities.

HOLZER: Ich danke Ihnen noch einmal sehr, für Ihren Vortrag und den Diskussionsrednern, sowie Herrn Dr. BREUER für seine Bemerkung, und darf damit den ersten Teil dieser Vormittagssitzung beenden.

Aktiver Transport von Aminosäuren*

Von

Erich Heinz

Institut für vegetative Physiologie, Frankfurt/M.

Mit 10 Textabbildungen

Die Fähigkeit, Aminosäuren anzureichern und zu transportieren, ist unter den lebenden Zellen und Geweben weit verbreitet. Ein Transport von Aminosäuren wurde u. a. beobachtet bei der Darmschleimhaut, beim Nierentubulus, beim Uterus, bei der Placenta, bei Muskelzellen, Nervenzellen, Blutkörperchen, Tumorzellen und außerdem bei zahlreichen Mikroorganismen und pflanzlichen Geweben. Besonders eingehend wurden in dieser Hinsicht die Darmschleimhaut, die Ehrlich-Maus-Ascites-Tumorzelle und verschiedene Mikroorganismen untersucht. Die Transportsysteme für Aminosäuren sind trotz der Verschiedenartigkeit der Zellen, in denen sie vorkommen, einander sehr ähnlich, vor allem in ihrer Spezifität, in ihrer Kinetik und in der Beeinflußbarkeit durch Hemmstoffe und Stimulantien. Grundsätzliche Befunde, die an bestimmten Zellarten erhoben wurden, gelten meist auch für andere Zellen und Gewebe.

Spezifität des Aminosäuretransportes

Die Spezifität des Aminosäuretransportes ist im Vergleich zu manchen Enzymreaktionen sehr breit. Viele Aminosäuren hemmen einander kompetitiv beim Transport. Daraus wird geschlossen, daß das einzelne Transportsystem nicht nur für eine bestimmte Aminosäure, sondern für eine ganze Gruppe oder Familie von Aminosäuren spezifisch ist. Es scheint mindestens drei voneinander unabhängige Transportsysteme zu geben: eins für neutrale, eins für basische und eins für saure Aminosäuren. Histidin findet sich in der Regel in der Familie der neutralen Aminosäuren, wogegen Cystin — zumindest im Nierentubulus — als Diaminosäure trans-

* Die Arbeit wurde unterstützt durch einen Grant (NSF-10812) der National Science Foundation, U.S.A.

portiert zu werden scheint[1]. Bei manchen Mikroorganismen ist die Spezifität etwas mehr differenziert: so z. B. konkurrieren im E. coli die verzweigten aliphatischen Aminosäuren miteinander, jedoch nicht mit den übrigen neutralen Aminosäuren[2]. Daneben sind einige unabhängige Transportsysteme für Peptide bekannt[3,4]. In der Regel wird das Peptid bereits während des Transportes oder bald darauf hydrolytisch gespalten.

Nicht alle Zellarten enthalten gleichzeitig alle drei Transportsysteme. Die Schleimhaut des Rattendarms transportiert z. B. *neutrale* Aminosäuren gegen das Konzentrationsgefälle, doch höchstwahrscheinlich keine *basischen* oder *sauren* Aminosäuren. Bei den sauren Aminosäuren ist möglicherweise der Sachverhalt durch Desaminierungen während des Transportes verschleiert[3]. Bei den Ehrlich-Ascites-Tumorzellen dagegen sind alle drei Transportsysteme für Aminosäuren weitgehend unabhängig voneinander wirksam[5,6]. Der Grad der Anreicherung ist auch innerhalb der gleichen Transportfamilie von Aminosäure zu Aminosäure verschieden.

Fast alle bisher beschriebenen Systeme für Aminosäuretransport sind deutlich, jedoch nicht absolut stereospezifisch. Die L-Konfigurationen werden in der Regel stark bevorzugt, obgleich auch die D-Konfigurationen — wenn auch schwächer — transportiert und dabei durch ihre optischen Antipoden kompetitiv gehemmt werden[3,7]. Soweit bisher bekannt ist, unterscheiden sich die optischen Antipoden voneinander durch ihre Affinität zum Transportsystem, d. h. durch die Michaelis-Konstante, nicht durch die Maximalgeschwindigkeit[3].

Zustand der Aminosäuren innerhalb der Zelle

Um entscheiden zu können, ob aktiver Transport vorliegt oder nicht, muß man die Konzentration der freien Aminosäuren zu beiden Seiten der transportierenden Membran kennen. Das ist kein Problem beim sog. „transcellularen" Transport z. B. beim isolierten Darm. Dagegen ist der Zustand des innerhalb von Zellen angereicherten Substrates oft schwer zu ermitteln. Die Tatsache, daß die intracelluläre Anreicherung von Aminosäuren vom Energiestoffwechsel der Zelle abhängt und daß der größte Teil der angereicherten Aminosäuren in wäßrigen Lösungen eiweißfrei extrahiert werden kann, schließt eine lose, endergonische Bindung der Aminosäuren an intracellulares Material nicht aus. Gegen eine derartige Bindung sprechen jedoch zahlreiche Befunde, von denen die wichtigsten kurz erwähnt werden sollen:

1. Die intracellulären Aminosäuren tauschen sich schnell und fast vollständig gegen extracelluläre Aminosäuren aus[8].

2. Die angereicherten Aminosäuren sind osmotisch aktiv, was an den Wasserverschiebungen bei erhöhter intracellulärer Anreicherung von Aminosäuren erkennbar ist[9].

3. Hemmung des Stoffwechsels beeinträchtigt nur den Influxkoeffizienten, jedoch kaum den Effluxkoeffizienten. Wenn bei Stoffwechselhemmung hauptsächlich intracelluläre Bindungen gelöst würden, wäre das Umgekehrte zu erwarten[8].

4. Vorherige Anreicherung von Ehrlich-Ascites-Tumorzellen mit Glykokoll oder analogen Aminosäuren steigert den Influx der betreffenden Aminosäure[10]. Auch hier wäre das Umgekehrte zu erwarten, wenn die Geschwindigkeit der Aufnahme von der Geschwindigkeit der intracellulären Bindung bestimmt würde.

Auch bei Mikroorganismen sind zahlreiche Phänomene beschrieben worden, die dafür sprechen, daß die angereicherten Aminosäuren im Innern der Zelle größtenteils frei gelöst sind. Ein großer Teil solcher Beobachtungen ist u. a. von MAGASANIK referiert worden[11].

Die Gesamtheit aller Befunde berechtigt also zu der Annahme, daß die intracellulären Aminosäuren durch einen aktiven Transportvorgang angereichert werden. Die kinetische Analyse der Aufnahme und des Austausches von Aminosäuren läßt darauf schließen, daß der Durchtritt der Aminosäure durch die transportierende Membran die Geschwindigkeit der Aufnahme bestimmt und daß Diffusionshindernisse im Innern der Zelle gegenüber der Zellmembran zu vernachlässigen sind[8]. Die Gleichgewichtsverteilung von nicht aktiv transportierten Substanzen ähnlicher Molekülgröße zwischen Zelle und Umgebung deutet darauf hin, daß die von der Zelle aufgenommenen Aminosäuren sich ziemlich gleichmäßig über den gesamten intracellulären Lösungsraum verteilen können. Es wurde ferner gezeigt, daß in den Ehrlich-Ascites-Zellen zumindest das Glykokoll während der Dauer der Experimente weder chemisch verändert noch in nennenswertem Umfang ein- oder abgebaut wird[8,12].

HALVORSON u. Mitarb. berichteten über Anhaltspunkte dafür, daß neben den frei austauschbaren Aminosäuren ("expandable pool") ein zweiter "pool" von ebenfalls extrahierbaren Aminosäuren besteht, die sich kaum gegen externe Aminosäuren austauschen und die hauptsächlich aus dem endogenen Stoffwechsel stammen ("internal pool")[13]. Die Aminosäuren des "internal pool" scheinen nicht den aktivierten Aminosäuren oder den Aminosäuren der löslichen RNS zu entsprechen. Die Aminosäuren des austauschbaren Pools sollen vor ihrem Einbau in Eiweiß zunächst in den "internal pool" übergeführt gewerden. Die Enzyme, welche die Aminosäuren aus dem "internal pool" in die Eiweißkörper einbauen, sollen wesentlich spezifischer als der aktive Transportmechanismus sein. HALVORSON fand verschiedene Anhaltspunkte dafür, daß Aminosäuren im extracellulären Raum auch direkt — d. h. unter Umgehung des austauschbaren Pools — in den "internal pool" gelangen und von da aus ins Eiweiß aufgenommen werden können.

Es ist anzunehmen, daß auch bei den Asciteszellen ein derartiger "internal pool" existiert. Er entspricht vermutlich den etwa 5—10% der Gesamt-Aminosäuren, die sich nicht ohne weiteres austauschen lassen[8]. Dieser "pool" ist also zu klein, um bei den kinetischen Untersuchungen des Transportmechanismus störend ins Gewicht zu fallen. Auch läßt sich bei ständigem Waschen der Asciteszellen nicht alles Glykokoll entfernen. Ob es sich bei diesem Rest-Glykokoll um den sog. "internal pool" handelt oder ob es dem endogenen Glykokoll entspricht, welches ebenso schnell neu gebildet wie ausgewaschen wird, läßt sich zur Zeit noch nicht entscheiden.

Kinetik des Aminosäuretransports

Wie bereits an anderen Stellen dargelegt wurde[6,8,10], zeigt die Transportkinetik von Aminosäuren alle für den aktiven Transport charakteristischen Merkmale wie Sättigungskinetik, kompetitive und nichtkompetitive Hemmbarkeit, Abhängigkeit vom Energiestoffwechsel der Zelle, einen hohen Temperaturkoeffizienten und eine definierbare, wenn auch breite Substratspezifität. Unter gleichbleibenden Bedingungen stellt sich in bezug auf den Aminosäuregehalt der Zelle ein stationärer Zustand ein, in welchem die betreffende Aminosäure ebenso schnell in die Zelle hineingepumpt wird, wie sie dieselbe verläßt. Die Aminosäure entweicht u. a. durch freie Diffusion, deren Geschwindigkeit von der Membranstruktur (Porenweite usw.) bestimmt wird. Wir haben die Geschwindigkeit dieses Umlaufs von Aufnahme und Abgabe während des stationären Zustandes bestimmt und dabei festgestellt, daß bei endogenem Stoffwechsel pro Molekül Sauerstoff maximal etwas über drei Glykokollmoleküle transportiert werden können[14]. Wenn man diese Vorgänge mit Hilfe von radioaktiven Isotopen in die unidirektionalen Fluxe zerlegt, so zeigt sich, daß die Fluxe etwa fünfmal größer sind, als nach den eben erwähnten Aufnahmevorgängen zu erwarten wäre[14]. Dies beruht darauf, daß neben aktivem Transport und freier Diffusion noch eine beträchtliche „Austauschdiffusion" stattfindet. Diese Austauschdiffusion kann nach der Definition USSINGs den Aminosäuregehalt der Zelle nicht verändern und ist vom Energiestoffwechsel unabhängig[15]. Wie bereits erwähnt, steigt der Influx mit der intracellulären Aminosäurekonzentration („Transkonzentration") erheblich an. Daraus ergibt sich, daß die Geschwindigkeit der Austauschdiffusion bei gleicher Außenkonzentration in hohem Maße von der intracellulären Konzentration der betreffenden Aminosäure abhängt, wie es für echte Austauschdiffusion charakteristisch sein sollte. Die chemische Spezifität der Austauschdiffusion stimmt weitgehend mit der des Transportes überein; d. h. die gleichen Aminosäuren, die die Aufnahme von Glykokoll kompetitiv hemmen, wie z. B. Sarkosin und Methionin, steigern den Glykokollinflux, wenn sie im Innern der Zelle angereichert sind, und umgekehrt (Hetero-Austauschdiffusion)[6]. Wir sehen daher Transport und Austauschdiffusion als Erscheinungsformen des gleichen Transportmechanismus an. In diesem Sinne ist die Größe der Austauschdiffusion ein Maß für die Reversibilität des Transportvorgangs. Wir halten das von uns beobachtete Austauschphänomen für wesensverwandt mit demjenigen, welches später von ROSENBERG und WILBRANDT unter der Bezeichnung "uphill transport induced by counter flow"[16] beschrieben worden ist.

Die kinetischen Befunde zeigen nahezu unwiderlegbar, daß die Aminosäure sich während ihres Transports vorübergehend an einen Überträger (Carrier) bindet, mit dessen Hilfe sie durch die „osmotische Schranke" geschleust wird. Aus der Gesamtheit der Beobachtungen haben wir folgendes kinetisches Modell abgeleitet (Abb. 1). Der Transport gegen das Konzentrationsgefälle setzt voraus, daß das Substrat vorzugsweise in einer Richtung übertragen wird. Dies wird durch die Annahme erklärt, daß der Überträger an der Innenseite der transportierenden Region katalytisch inaktiviert wird und dadurch seine Affinität zur Aminosäure ganz oder zum größten Teil verliert. Das Phänomen der Austauschdiffusion setzt bei diesem Modell zwangsläufig voraus, daß die Reaktionsschritte 1—4 reversibel sind und daß diese Reaktionen wesentlich schneller ablaufen als diejenigen, durch welche der Carrier zurückgeführt bzw. reaktiviert wird. Die Transportgeschwindigkeit wird also durch eine der letztgenannten Reaktionen bestimmt.

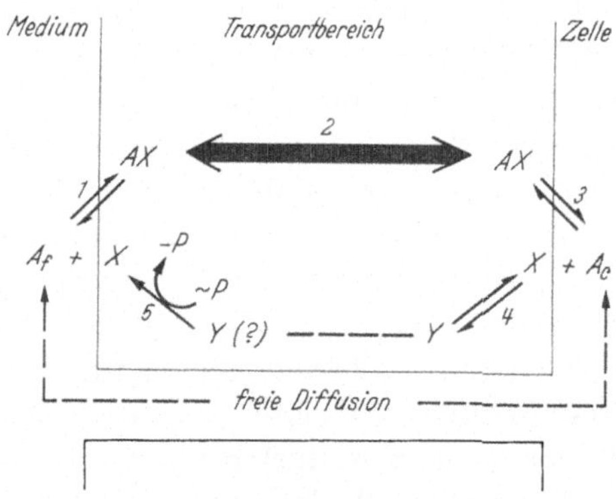

Abb. 1. *Modell für aktiven Transport und Austauschdiffusion von Aminosäuren.* A_f bedeutet das extracelluläre, A_c das intracelluläre Transportsubstrat. Die aktive Form des Carriers ist durch X, die inaktive Form durch Y gekennzeichnet. AX ist der Carrier-Substrat-Komplex. $\sim P$ und $-P$ sollen ganz allgemein Ausgangs- und Endprodukte der energieliefernden Stoffwechselreaktion darstellen (nach E. HEINZ und P. M. WALSH[6])

Der Reaktionsschritt 2 ist hier nur der Anschaulichkeit halber als translatorische Bewegung des mutmaßlichen Carrier-Substrat-Komplexes dargestellt. Ohne Änderung des kinetischen Modells könnte dieser Reaktionsschritt auch eine Rotationsbewegung, eine Pendelbewegung oder eine sonstige statistische Umordnung von membraneigenen Bestandteilen mit entsprechender Verschiebung der aminosäurebindenden (aktiven) Gruppen verkörpern[17]. Von MITCHELL wird neuerdings eine Theorie vertreten, nach der der Carrier sich beim Transport überhaupt nicht zu bewegen braucht[18]. Er hebt hervor, daß in vielen enzymatischen Reaktionen Substrate während der Bindung an ein Enzym sich relativ zu diesem Enzym bewegen müssen, um mit dem Partnersubstrat reagieren zu können. Eine derartige Bewegung müßte entlang der Enzymoberfläche oder durch das Enzym hindurch erfolgen. Sie tritt makroskopisch als Transport in Erscheinung, wenn diese Enzyme anisotrop, d. h. innerhalb einer Diffusionsschranke gleichartig orientiert sind, indem sie etwa auf der einen Seite nur für den einen und auf der anderen Seite nur für den anderen Reaktionspartner zugänglich sind: das Enzym wird zur ,,Translokase". Auf diese Weise wird ein Transportvorgang als Sonderfall einer ,,orthodoxen" enzymatischen Gruppenübertragungsreaktion interpretiert. Diese Gedankengänge haben uns dazu angeregt, unser Modell entsprechend zu modifizieren, wodurch es gewissermaßen zu einem zweidimensionalen Gebilde verflacht wird. Viele derartige Modelle sind möglich, die einen aktiven Transport erklären, aber nur wenige, die außerdem mit dem Phänomen der Austauschdiffusion bzw. dem "counter flow"-Effekt zu vereinbaren sind. Das vorliegende Modell scheint dieser Forderung zu genügen (Abb. 2). Von der Innenseite der Zelle her reagiert ein aktivierter Acceptor XP mit der ,,Translokase" E unter Bildung eines Enzym-Acceptor-Komplexes EX. In dieser Form kann X mit der von der Außenseite an die ,,Translokase" herankommenden Aminosäure reagieren unter Bildung eines neuen Komplexes AX. Dabei wird AX von E abgelöst, und durch Spaltung von AX an irgendeiner Stelle der Zelle kommt es dann zur Anreicherung von A, deren Ausmaß durch das chemische Gruppenpotential von AX gegeben ist. Das Phänomen der Austauschdiffusion verlangt u. a., daß alle diese Vorgänge reversibel sind.

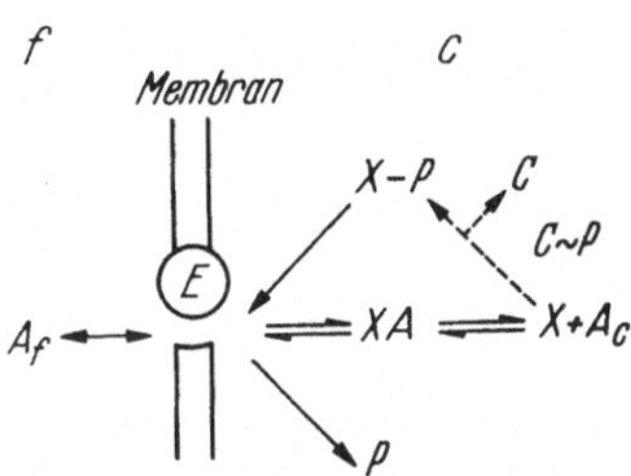

Abb. 2. *Translokase"-Modell, entwickelt nach der Theorie von* P. MITCHELL[18]. A_f und A_c = extra- und intracelluläres Substrat; E = Translokase; X = intracellulärer Acceptor; $C \sim P$ = energiereicher Reaktionspartner für die Aktivierung von X

Veränderungen am Substrat

Um Aufschluß über die Art der Bindung zwischen Transportsubstrat und Überträger zu erhalten, kann man das Transportsubstrat in charakteristischer Weise modifizieren und dabei die Voraussetzungen für seine Affinität zum Transportsystem (Transportaffinität) ermitteln. Die ausgedehntesten Untersuchungen dieser Art wurden bisher bei den Ehrlich-Ascites-Tumorzellen in bezug auf den Transportmechanismus für neutrale Aminosäuren durchgeführt[5,7,19]. Die Gesamtheit der Ergebnisse läßt sich etwa in folgendem Bilde zusammenfassen (Abb. 3). Die unentbehrlichen Bestandteile des Moleküls sind fettgedruckt. Sowohl die Carboxylgruppe als auch die Aminogruppe müssen ionisiert sein. Dementsprechend darf z. B. die Carboxylgruppe nicht verestert, amidiert oder reduziert werden. Sie kann außerdem nicht durch eine Sulfonsäuregruppe ersetzt werden. Die Aminosäuregruppe kann nicht

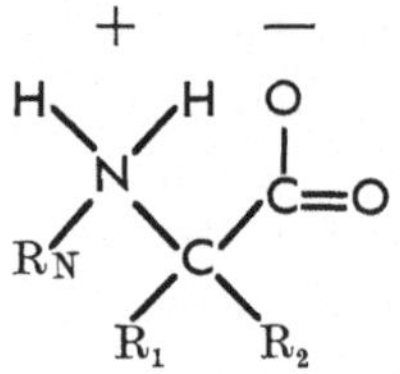

+ —

H H O

N C=O

R_N C

R_1 R_2

R_N	R_1	R^2
—H	—H	—H
$—CH_3$	$—C_nH_{2n}$(gestreckt)	$—CH_3$
	$—C_nH_{2n}—OH$	
	$—C_nH_{n2}—SH$	
	$—C_nH_{2n}—S—CH_3$	
	$—C_nH_{2n}—CO—NH_2$	
	$—C_nH_{2n}$—aryl	

Abb. 3. *Zur Spezifität des Transportmechanismus für neutrale Aminosäuren*

acyliert oder amidiniert werden. Sie verträgt ohne weiteres eine Methylgruppe. Dagegen wird die Affinität durch Einführung einer Äthylgruppe oder mehrerer Methylgruppen völlig aufgehoben. Die optische Spezifität läßt erwarten, daß außer diesen beiden Gruppen noch eine dritte sich an der Reaktion mit dem Transport-Carrier beteiligt. Nach unserer Meinung stellt eine Alkylgruppe in der α_1-Stellung diese dritte Gruppe dar, da die α-Aminoisobuttersäure und das L-Alanin eine wesentlich größere Transportaffinität zeigen als Glykokoll und D-Alanin. Die Länge dieser Alkylgruppe spielt offenbar keine Rolle, dagegen wirkt sich Verzweigung ungünstig aus. OH-, SH- und SCH_3-Gruppen beeinträchtigen die Transportaffinität nicht. Ionisierung der dritten Gruppe hebt die Affinität für dieses betreffende Transportsystem auf. Wie bereits erwähnt, werden die sauren bzw. basischen Aminosäuren von einem unabhängigen Transportsystem erfaßt. Entionisierung dieser Gruppe z. B. durch Amidierung stellt die Transportaffinität wieder her[7].

Die eben erwähnten Befunde stimmen mit der Annahme überein, daß das Substrat mit dem Überträger an drei Stellen reagiert. Über die Art der Bindung ist noch nichts bekannt.

Veränderungen am Trägersystem

Es ist verschiedentlich versucht worden, den hypothetischen Träger (Carrier) bzw. die maßgebenden aktiven Gruppen spezifisch zu blockieren oder zu modifizieren. Von den entsprechenden Verfahren, die in der Enzymchemie verwendet werden, sind nur wenige für die lebenden Zellen unschädlich. Ein Erfolg in dieser Richtung wurde von QUASTEL u. Mitarb. berichtet[20]. In Gegenwart eines spezifisch gegen Ehrlich-Maus-Ascites-Tumorzellen gerichteten Antiserums war der Glykokolltransport stark beeinträchtigt, ohne daß eine andere Störung der Zelle zu erkennen war. Dieses Ergebnis scheint vor allem anzudeuten, daß der Überträger ganz oder teilweise aus Eiweiß besteht.

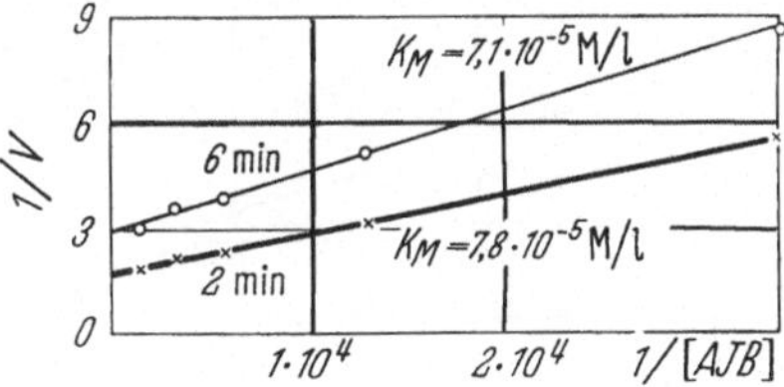

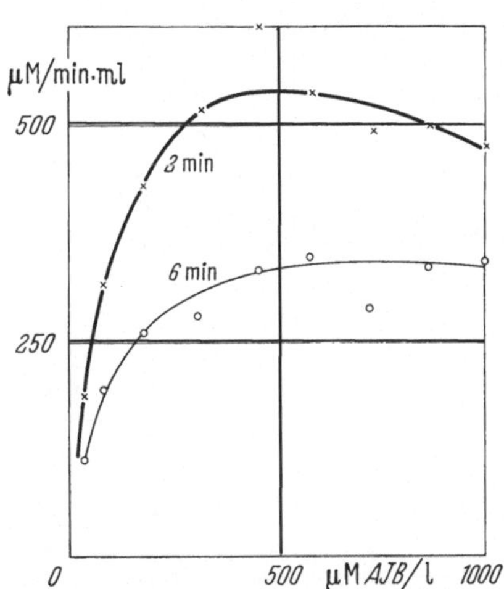

Abb. 4. *Aufnahme von radioaktiver α-Aminoisobuttersäure (AIB) durch Streptomyces hydrogenans.* Unten: Abhängigkeit von der extracellulären AIB-Konzentration (+ nach 2 min, o nach 6 min). Oben: Auswertung der gleichen Daten nach LINEWEAVER und BURK (nach J. HÜBENER, in Vorbereitung)

Weitere Hinweise auf die enzymatische Natur des Aminosäuretransportes hat kürzlich in unserem Institut HÜBENER mit dem Strahlenpilz Streptomyces hydrogenans geliefert. Dieser Mikroorganismus reichert α-Aminoisobuttersäure (AIB) bis zu tausendfach an (Abb. 4). Die Kinetik ist ähnlich wie bei den Ehrlich-Ascites-Tumorzellen, nur scheint die Affinität der α-Aminoisobuttersäure zu dem Transportsystem des Streptomyces über sechzigmal größer zu sein als die des Glykokolls zu dem entsprechenden System der Tumorzellen. HÜBENER fand Anhaltspunkte dafür, daß das transportierende System für Aminosäuren dieses Aktinomyces adaptiv (induktiv) gebildet wird. Das steht im Gegensatz zu anderen Mikroorganismen, deren Aminosäuretransport nicht induzierbar ist. Setzt man nämlich dem im synthetischen Nährmedium wachsenden Pilz für einige Stunden ein Aminosäuregemisch zu und wäscht ihn anschließend vom Medium frei, so reichert er AIB etwa zehnmal stärker als die entsprechenden Kontrollen an (Abbildung 5). Die kinetische Analyse ergab, daß hierfür ein vermehrter Influx bei fast unverändertem Efflux verantwortlich ist. Diese Steigerung des Influxes beruht nicht auf einem Gegenstrom der vorher angereicherten Aminosäuren (Austauscheffekt), da Vorbeladung der Zellen mit AIB den Influx radioaktiver AIB nicht beeinflußt. Auch Vorbehandlung mit Alanin scheint den AIB-Transport zu induzieren.

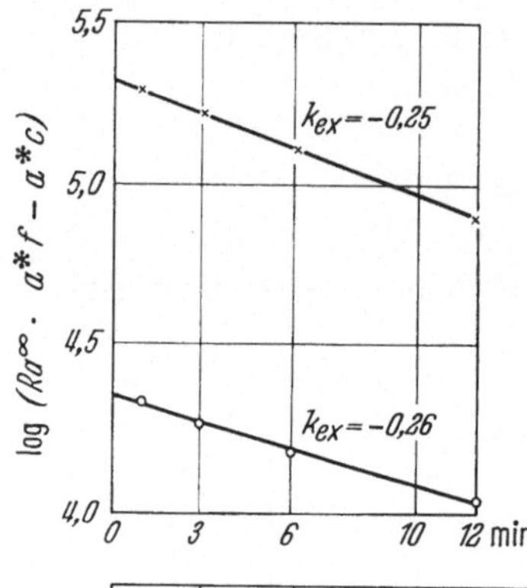

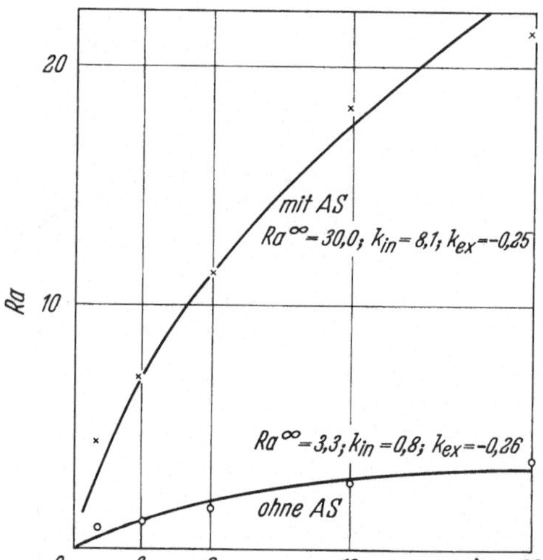

Abb. 5. *Aufnahme von radioaktiver AIB durch Streptomyces hydrogenans nach Vorbehandlung mit Aminosäuregemisch (AS).* Unten: Verteilungsquotient der AIB zwischen intra- und extracellulärer Flüssigkeit in Abhängigkeit von der Zeit. Oben: Auswertung der Ergebnisse zur Ermittlung der Flux-Koeffizienten (K_{ex} und K_{in}) (nach E. HEINZ[8] und nach J. HÜBENER, in Vorbereitung)

Es kann z. Z. noch nicht ausgeschlossen werden, daß die zugesetzten Aminosäuren, statt zu induzieren, essentiell zum Aufbau des transportierenden Systems notwendig sind. Interessant in diesem Zusammenhang ist, daß Zugabe von Äthionin zu den Mikroorganismen nach etwa sechs Stunden den Transportmechanismus stark beeinträchtigt. Diese Beeinträchtigung kann verhindert werden, indem man mit dem Äthionin gleichzeitig einen Überschuß von Methionin verabfolgt (Abb. 6). Die Induktionsversuche werden dadurch erschwert, daß stets geringe Aminosäuremengen vorhanden sind, die wahrscheinlich ein relativ hohes konstitutives Transportsystem unterhalten.

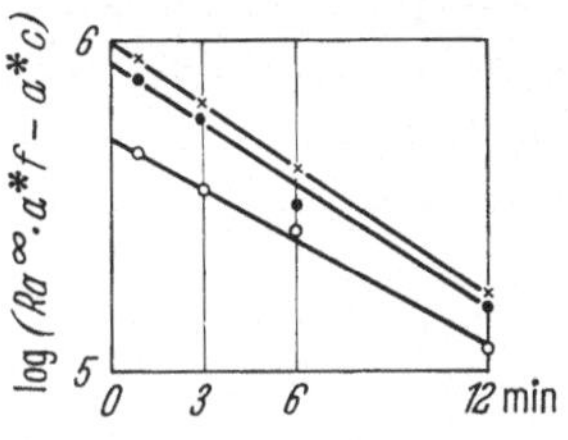

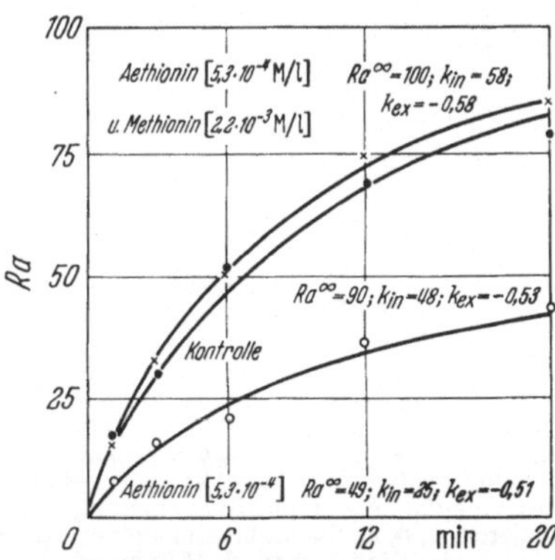

Abb. 6. *Einfluß von Äthionin und Äthionin + Methionin auf die zeitliche Aufnahme von radioaktiver AIB durch Streptomyces hydrogenans nach Vorbehandlung mit Aminosäuregemisch.* Leere Kreise: nach Vorbehandlung mit Äthionin; Kreuze: nach Vorbehandlung mit Äthionin + Methionin. Ordinate wie in Abb. 5. Oben: Auswertung der Flux-Koeffizienten wie in Abb. 5 (nach J. HÜBENER, in Vorbereitung)

Ähnlich wie für Enzyme ist auch das Transportsystem für Aminosäuren stark p_H-abhängig[5]. KROMPHARDT in unserem Laboratorium hat den Effekt der H-Ionen eingehend untersucht und dabei zunächst festgestellt, daß dieser Effekt sofort nach Zugabe der H-Ionen voll einsetzt und durch Wiederherstellung des normalen p_H sogleich wieder aufgehoben wird. Er ist also reversibel und wirkt auf das Transportsystem höchstwahrscheinlich nicht über den Stoffwechsel, sondern direkt ein. Die Abhängigkeit der Hemmung vom p_H der Lösung entspricht einer Säure-Basen-Titrationskurve, doch bleibt selbst bei den höchsten H-Ionenkonzentrationen ein p_H-unabhängiger „Resttransport" bestehen. Diese Beobachtung läßt vermuten, daß der hemmbare Teil des Transportes von einer basischen Gruppe abhängig ist, deren Säure-pK bei etwa 7 liegt. Die Hemmung durch H-Ionen ist nicht kompetitiv, denn der scheinbare pK wird selbst bei starker Erhöhung der Glykokollkonzentration (von 0,6 auf 20 mmolar) nicht verschoben (Abb. 7). Es ist daher unwahrscheinlich, daß sich die betreffende Gruppe direkt an der Bindung der Aminosäure als Haftpunkt beteiligt. Der mutmaßliche pK dieser Gruppe läßt an den Imidazolring des Histidins denken. Wir haben versucht, diesen Imidazolring durch Diazotierung anzugreifen. Es war uns bisher jedoch nicht möglich, den Glykokolltransport eindeutig zu hemmen, ohne gleichzeitig auch andere Funktionen der Zelle wie Atmung und Elektrolyttransport zu beeinträchtigen.

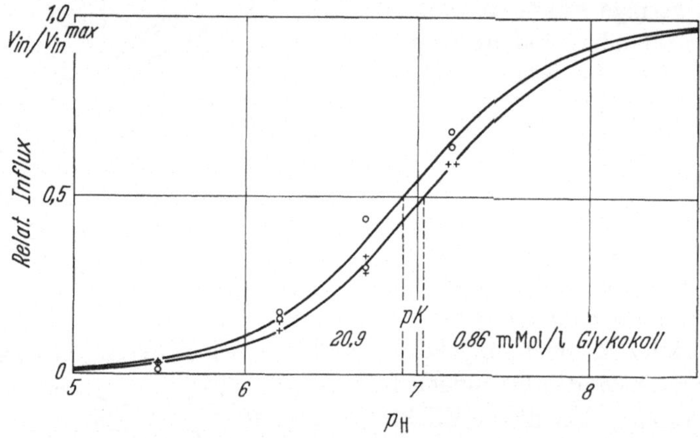

Abb. 7. *Einfluß der H-Ionen auf den Glykokollinflux in Ehrlich-Ascites-Tumorzellen bei verschiedenen extracellulären Glykokollkonzentrationen.* Die relativen Flux-Werte sind um eine kleine p_H-unabhängige Effluxkomponente vermindert. Der Maximal-Influx ist durch Extrapolation erhalten worden. Die gestrichelten Geraden geben die Wendepunkte bei den extracellulären Glykokollkonzentrationen von 0,86 mMol pro l (rechts) und von 20,9 mMol pro l (links) an (nach H. KROMPHARDT und E. HEINZ, in Vorbereitung)

Nach P. MITCHELL handelt es sich beim aktiven Transport — wie bereits angedeutet — um eine enzymatische Gruppenübertragung. So wird z. B. beim Micrococcus lysodeicticus die Bernsteinsäure nicht als solche, sondern als eine vermutlich an Coenzym A gebundene Succinylgruppe transportiert[21]. MITCHELL nimmt an, daß in analoger Weise die Aminosäuren nach Carboxylaktivierung als Aminoacylgruppen durch die Zellmembran treten[22]. Dagegen spricht, daß die Aminosäuren bei Aktivierung mit ATP ein Sauerstoffatom mit der Adenylsäure austauschen, während der Durchtritt der Aminosäure durch die Zellmembran höchstwahrscheinlich ohne Sauerstoff-Austausch verläuft[23,24]. Dieser Einwand ist schwerwiegend, aber nicht absolut. KELLERMAN hat kürzlich eine enzymatische Reaktion beschrieben, bei welcher eine carboxyl-aktivierte Verbindung — Benzoyl-AMP — in einer Weise gespalten wird, daß der Sauerstoff der Anhydridbindung beim Benzoesäurerest verbleibt[25]. Wenn nämlich das Substrat mit dem Enzym verbunden bleibt, erfolgt die Spaltung zwischen O und P, während bei der nicht enzymatischen hydrolytischen Spaltung die Bindung zwischen C und O gelöst wird. Auf diese Weise kann eine Carboxylgruppe eine Säureanhydridbindung durchlaufen, ohne dabei ein Sauerstoffatom auszutauschen. Obgleich eine derartige Reaktion anscheinend für Aminosäuren noch nicht beschrieben worden ist, läßt sie sich für den Transport nicht mit Sicherheit ausschließen.

Es bleibt die Möglichkeit, daß die Aminosäure während des Transportes an der Aminogruppe aktiviert wird. Diese Frage wurde besonders ausgiebig von CHRISTENSEN im Zusammenhang mit der von ihm entdeckten Wirkung des Pyridoxals auf den Aminosäuretransport diskutiert. Das Pyridoxal sollte als Cofaktor des Carriers fungieren und dabei mit der Aminogruppe vorübergehend eine Schiffsche Base bilden. In die gleiche Richtung deuten einige andere Befunde. Der Lactobacillus arabinosus ist bei spezifischem Mangel an Vitamin B_6 in seinem Vermögen, Glutaminsäure anzureichern, sehr stark beeinträchtigt[26]. Allerdings ließ sich diese Hemmung durch Zugabe von Pyridoxal oder Pyridoxalphosphat allein nicht beheben. Ähnliche Beobachtungen wurden von SNELL mit Lactobacillus casei in bezug auf Glykokoll- und Alanintransport gemacht[4]. Die intestinale Resorption von DL-Alanin (FRIDHANDLER)[27] und von Methionin (JAKOBS u. Mitarb.)[28] wurde durch das Antivitamin Desoxypyridoxin gehemmt. Diese Hemmung konnte durch Verabfolgung von Pyridoxin gemildert werden. Pyridoxin, Pyridoxalphosphat und Pyridoxal schützten ferner die intestinale Resorption von L-Tyrosin gegen Vergiftung durch 2,4-Dinitrophenol[29,30].

Gegen die Annahme einer Schiffschen Base spricht zunächst, daß auch Sarkosin und Prolin, deren sekundäre Aminogruppen nur eine geringe, nur bei alkalischer Reaktion ausgeprägte Tendenz zur Schiffschen Basen-Bildung haben, die gleiche Affinität zum Carrier besitzen wie das Glykokoll. Auch konnte CHRISTENSEN den Transport des Glykokolls durch carbonylblockierende Reagentien wie Semicarbazid und Isonicotinsäurehydrazid nicht hemmen. Er fand außerdem, daß auch Oestradioldisulfat und Diäthylstilboestroldisulfat trotz der fehlenden Carbonylgruppe die Anreicherung des Glykokolls in Ehrlich-Ascites-Tumorzellen steigern[31].

Wenn das Pyridoxal direkt am Transport beteiligt wäre, so müßte sein Effekt auf die Anreicherung des Glykokolls mit einer Steigerung des Influxes verbunden sein. Wir haben jedoch in keinem Falle eine eindeutige Wirkung von Pyridoxal in dieser Richtung beobachten können[26]. Im gleichen Sinne sprechen auch die

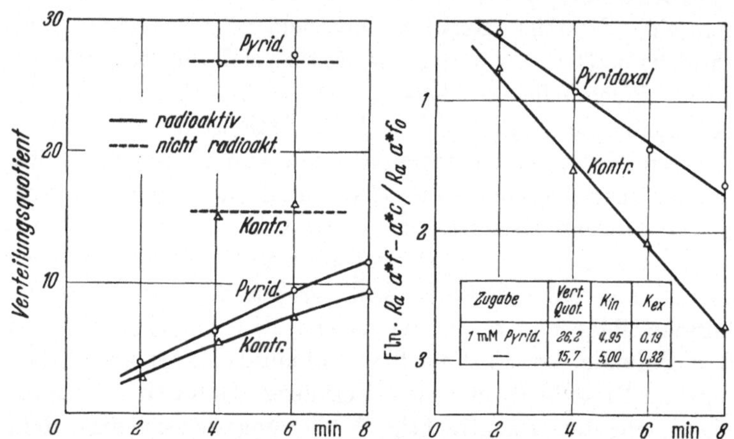

Abb. 8. *Zeitliche Aufnahme von radioaktivem Glykokoll durch Ehrlich-Ascites-Tumorzellen nach Vorbehandlung mit 1 mMol Pyridoxal.* Links: Die gestrichelten Linien beziehen sich auf die entsprechenden steady state-Konzentrationen des gesamten Glykokolls. Rechts: Auswertung der Aufnahmedaten zur Ermittlung der Flux-Koeffizienten wie in Abb. 5. und 6 (nach J. BITTNER und E. HEINZ, in Vorbereitung)

Befunde von HOLDEN u. Mitarb.[33], nach denen bei L. arabinosus der Vitamin B_6-Mangel nur die endgültige Verteilung von Glutaminsäure, nicht aber die Anfangsgeschwindigkeit der Aufnahme dieser Aminosäure beeinflußt. Sie kamen zu dem Schluß, daß bei Pyridoxalmangel die Zellwand geschädigt wird, der Transport jedoch nicht unmittelbar betroffen ist. BITTNER in unserem Laboratorium hat den steigenden Einfluß von 1 mMol Pyridoxal auf die Glykokollanreicherung bestätigen können. Sie hat ferner diese Anreicherung unter Pyridoxaleinwirkung kinetisch in die unidirektionalen Fluxe zerlegt und dabei eindeutig festgestellt, daß die Anreicherung durch Pyridoxal auf einer Verminderung des Effluxkoeffizienten beruht, während der Influxkoeffizient in Übereinstimmung mit unseren früheren Resultaten durch Pyridoxal nicht signifikant verändert wird (Abb. 8). Dieser Befund läßt vermuten, daß entweder das unter der Wirkung von Pyridoxal vermehrt angereicherte intracelluläre Glykokoll vermindert austauschfähig ist oder daß entsprechend der Annahme von CHRISTENSEN der passive Ausstrom von Glykokoll durch Abdichtung der Membran verhindert wird[31]. Weitere Untersuchungen zur Klärung dieses Sachverhaltes sind erforderlich.

Die bisher genannten Befunde scheinen anzudeuten, daß die Aminosäuren während des Transportes keine kovalente Bindung mit dem Überträger oder Enzym eingehen. Höchstwahrscheinlich handelt es sich nur um elektrostatische Bindungen zwischen den beiden ionisierten Gruppen und entsprechenden aktiven Gruppen des Acceptors. Über die Funktion der dritten Gruppe bei dieser Bindung kann z. Z. noch nichts gesagt werden.

Energiebedarf des Transportes

Bei Abwesenheit eines exogenen Substrats wird der Glykokolltransport durch Anaerobiose gehemmt[14]. Zugabe von Glucose hebt diese Hemmung auf[34]. Aus der Hemmbarkeit des Transports durch 2,4-Dinitrophenol wird auf die wahrscheinliche Beteiligung energiereicher Phosphate geschlossen. Über die energetische Kopplung zwischen Transport und energieliefernder Reaktion ist noch nichts bekannt. Grundsätzlich käme jeder der in den Modellen postulierten Reaktionsschritte für eine derartige Kopplung in Frage. Wir halten es für wahrscheinlicher, daß nur die Reaktionen der Rückführung und der Reaktivierung des Überträgers (Carriers) energetisch gekoppelt sind, wie wir es hypothetisch auf unserem Modell dargestellt haben. JAQUEZ kommt auf Grund einer eingehenden mathematischen Analyse der vorliegenden kinetischen Befunde mit und ohne Stoffwechselhemmung zu dem gleichen Ergebnis[35].

Eine neuartige Hypothese über die energetische Kopplung wurde von CHRISTENSEN u. Mitarb. vorgeschlagen[36]. Er beobachtete, daß die Anreicherung von Glykokoll in Ehrlich-Ascites-Tumorzellen vermindert wird, wenn man entweder das extracelluläre Kalium erhöht oder das intracelluläre vermindert. Er sah darin eine Abhängigkeit des Glykokolltransportes von dem Konzentrationsgefälle des Kaliums und schloß daraus, daß die bei dem Austausch des Kaliums frei werdende Energie zur Reaktivierung des Glykokoll-Carriers verwendet würde[33]. Demnach bezöge der Glykokolltransport seine Energie nicht unmittelbar vom Stoffwechsel, sondern über den Umweg des Kaliumtransportes (Satelliten-Transport). Strenggenommen handelt es sich dabei um eine Art Hetero-Austauschdiffusion zwischen Kalium und Glykokoll. In früheren Untersuchungen hatten wir mit PATLAK den minimalen Energiebedarf des reinen Glykokoll-Transportmechanismus mit etwa 2000 Calorien pro Mol transportierten Glykokolls errechnet[37]. Dieser Energiebedarf stimmt in etwa mit der elektrochemischen Potentialdifferenz zwischen innerem und äußerem Kalium überein. Auch die Fluxe des Kaliums sind von ähnlicher Größenordnung wie die des Glykokolls[38]. Gegen die plausible Hypothese von CHRISTENSEN lassen sich also weder von thermodynamischer noch von kinetischer Seite her grundsätzliche Einwände erheben. KROMPHARDT und GROBECKER haben diese Hypothese in unserem Laboratorium

Tabelle 1. *Intrazellulärer K^+- und Glycin-Influx*

Intrazell. $[Na^+]$ (mVal/L)	Intrazell. $[K^+]$ (mVal/L)	Glycin-Influx ml/g/2 min
61	168	9,7
87	182	10,1
167	74	9,8
174	71	9,7

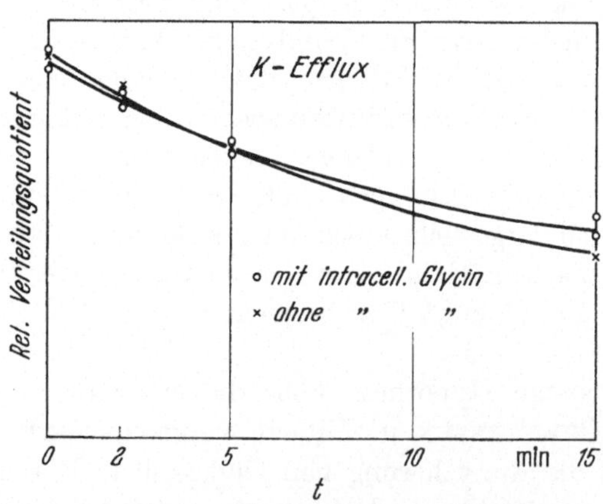

Abb. 9. *Austritt von ^{42}K auf Ehrlich-Ascites-Tumorzellen unter dem Einfluß von intracellulärem Glykokoll.* Kreise: nach Vorbeladung der Zellen mit Glykokoll; Kreuze: ohne Vorbeladung mit Glykokoll (nach H. GROBECKER, H. KROMPHARDT und E. HEINZ, in Vorbereitung)

untersucht. Sie fanden keine eindeutige Verminderung des Glykokoll-Influxes bei starker Verminderung des intracellulären Kaliums

(Tab. 1). Auch wäre nach der Christensenschen Hypothese zu erwarten, daß der Efflux des Kaliums durch starke Erhöhung des intracellulären Glykokolls vermindert würde. Eine derartige Verminderung wurde in unserem Laboratorium nicht beobachtet (Abb. 9). HEMPLING kam bei entsprechenden Versuchen zu ähnlichen Ergebnissen[39]. Die genannten experimentellen Befunde scheinen also die Hypothese einer direkten Kopplung zwischen

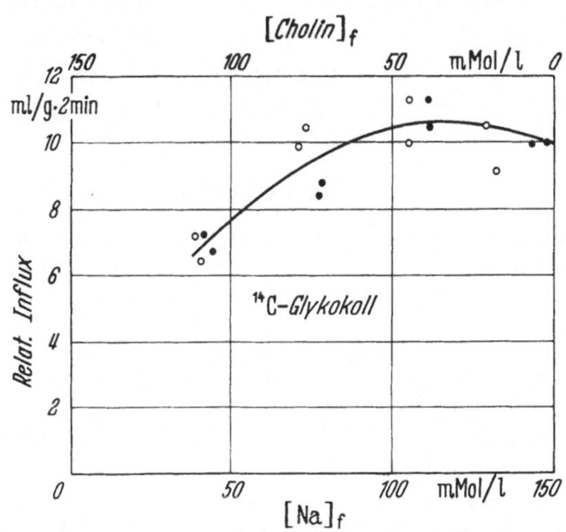

Abb. 10. *Abhängigkeit des Glykokollinfluxes in Ehrlich-Ascites-Tumorzellen von der extracellulären Natriumkonzentration.* Die extracelluläre Kaliumkonzentration war in allen Fällen 6 mMol pro l. Die Kationendefizite wurden durch Cholin ausgeglichen (nach H. GROBECKER, H. KROMPHARDT und E. HEINZ, in Vorbereitung)

Kalium-Ausstrom und Glykokolltransport zu widerlegen. Es ist daher wahrscheinlicher, daß der Glykokolltransport direkt mit einer energieliefernden Stoffwechselreaktion gekoppelt ist.

Es bleibt noch die Frage zu klären, ob Glykokolltransport und Kaliumhaushalt der Ehrlich-Zelle auf eine andere Weise miteinander verknüpft sind. HEMPLING fand kürzlich, daß Pyridoxal beide Transportsysteme — und zwar in entgegengesetzter Weise — beeinflußt[40]. Von CHRISTENSEN war schon früher berichtet worden, daß Inkubation dieser Zellen bei erhöhten Kaliumkonzentrationen die Akkumulation von Glykokoll stark beeinträchtigt[5]. Eine kurzfristige Einwirkung (2 min) erhöhter Kaliumkonzentrationen auf diese Zellen beeinträchtigt nach Untersuchungen von KROMPHARDT und GROBECKER den Glykokollinflux wesentlich schwächer.

Doch handelt es sich hierbei höchstwahrscheinlich um einen Scheineffekt, denn Kalium-Hemmung bleibt nahezu aus, wenn das extracelluläre Natrium gleichzeitig — unter Verwendung von Cholinionen — konstant gehalten wird. Auf der anderen Seite sinkt der Glykokollinflux, wenn die Natriumkonzentration unter einen kritischen Wert abfällt (Abb. 10). Vorausgesetzt, daß die Cholinionen sich in diesem Falle indifferent verhalten, sprechen die Befunde für einen Einfluß der extracellulären Natriumkonzentration auf den Glykokolltransport. Sie erinnern an diejenigen von CSÁKY u. a., nach denen der Zuckertransport durch die Darmschleimhaut von der Natriumkonzentration im Darmlumen abhängt[41].

Literaturverzeichnis

1 DENT, C. E., B. SENIOR and J. M. WALSH: J. clin. Invest. **33**, 1216 (1954).
2 COHEN, E. E., and H. V. RICKENBERG: Ann. Inst. Pasteur **91**, 693 (1956).
3 SMYTH, D. H.: Membrane Transport and Metabolism. Symp. Prag 1960 (im Druck).
4 LEACH, F. R., and E. E. SNELL: J. biol. Chem. **235**, 3523 (1960).
5 CHRISTENSEN, H. N., T. R. RIGGS, H. FISCHER and I. M. PALATINE: J. biol. Chem. **198**, 1 (1952).
6 HEINZ, E., and P. M. WALSH: J. biol. Chem. **233**, 1488 (1958).
7 PAINE, C. M., and E. HEINZ: J. biol. Chem. **235**, 1080 (1960).
8 HEINZ, E.: J. biol. Chem. **225**, 305 (1957).
9 CHRISTENSEN, H. N., T. R. RIGGS, H. FISCHER and I. M. PALATINE: J. biol. Chem. **198**, 17 (1952).
10 HEINZ, E.: J. biol. Chem. **211**, 781 (1954).
11 MAGASANIK, B.: Ann. Rev. Microbiol. **1957**, 232.
12 HEINZ, E.: Abstr. 129th Nat. Meet. Am. Chem. Soc., Div. of Biochem. Dallas (1956).
13 HALVORSON, H. O., and D. B. COWIE: Membrane Transport and Metabolism. Symp. Prag 1960 (im Druck).
14 HEINZ, E., and H. M. MARIANI: J. biol. Chem. **228**, 97 (1957).
15 USSING, H. H.: Physiol. Rev. **29**, 127 (1949).
16 ROSENBERG, T., and W. WILBRANDT: J. Gen. Physiol. **41**, 289 (1958).
17 PATLAK, C. S.: Bull. Math. Biophysics **19**, 209 (1957).
18 MITCHELL, P.: Proc. roy. Phys. Soc. Edinb. **27**, 61 (1958).
19 RIGGS, T. R., B. H. COYNE and H. N. CHRISTENSEN: J. biol. Chem. **209**, 395 (1954).
20 QUASTEL, J. H.: Membrane Transport and Metabolism. Symp. Prag 1960 (im Druck).
21 MITCHELL, P., and J. MOYLE: Biochem. J. **72**, 21 P (1959).
22 MITCHELL, P., and J. MOYLE: Proc. Roy. Phys. Soc. Edinb. **28**, 19 (1959).
23 BOYER, P. D., and M. P. STULBERG: Proc. Nat. Acad. Sci. (Wash.) **44**, 92 (1958).

24 CHRISTENSEN, H. N., H. M. PARKER and T. R. RIGGS: J. biol. Chem. **233**, 1485 (1958).
25 KELLERMAN, G. M.: J. biol. Chem. **231**, 427 (1958)
26 HOLDEN, J. T.: J. biol. Chem. **234**, 872 (1959).
27 FRIDHANDLER, L., and J. H. QUASTEL: Arch. Biochem. **56**, 424 (1955).
28 JACOBS, F. A., and R. S. L. HILLMAN: J. biol. Chem. **232**, 451 (1958).
29 JACOBS, F. A., L. J. COEN and R. S. L. HILLMAN: J. biol. Chem. **235**, 1375 (1960).
30 JACOBS, F. A., R. C. FLAA and W. F. BELK: J. biol. Chem. **235**, 3224 (1960).
31 CHRISTENSEN, H. N.: Membrane Transport and Metabolism. Symp. Prag 1960 (im Druck).
32 HEINZ, E., and C. M. PAINE: unveröffentlicht.
33 HOLDEN, J. T.: Biochim. biophys. Acta **33**, 581 (1959).
34 JOHNSTONE, R. M.: Canad. J. Biochem. **37**, 589 (1959).
35 JAQUEZ, J. A.: Proc. Nat. Acad. Sci. (Wash.) **47**, 153 (1961).
36 RIGGS, T. R., L. M. WALKER and H. N. CHRISTENSEN: J. biol. Chem. **233**, 1479 (1958).
37 PATLAK, C. S.: Biochim. biophys. Acta **44**, 324 (1960).
38 HEINZ, E., and H. A. MARIANI: Abstr. 132nd Nat. Meet. Am. Chem. Soc. Div. of Biol. Chem. p. 560. New York 1957.
39 HEMPLING, H. G.: Physiologist **3**, 78 (1960).
40 HEMPLING, H. G.: Fed. Proc. **20**, 139 (1961).
41 CSÁKY, T. Z., and M. THALE: J. Physiol. (Lond.) **151**, 59 (1960).
42 CRANE, R. K.: Membrane Transport and Metabolism. Symp. Prag 1960 (im Druck).

Diskussion

Diskussionsleiter: NETTER, *Kiel*

Ich danke Herrn HEINZ für seinen Vortrag und eröffne die Diskussion.

KARLSON (München): Worin liegt der Vorteil des „Transportase-Modells" gegenüber dem der Permease? Ich sehe eine Schwierigkeit bezüglich der Dimension. Ein Enzymmolekül mit etwa 50 Å Durchmesser ist um 1 bis 2 Größenordnungen kleiner als die Membran, kann also nur über eine monomolekulare Strecke transportieren. — Als Alternative kann man eine ausgedehnte Enzymfläche von Membrangröße annehmen, an der die Aminosäuren entlang kriechen. Es ist aber nicht einzusehen, weshalb das schneller gehen soll als Diffusion.

HEINZ: Auch meiner Ansicht nach ist dieses System dem vorherigen nicht unbedingt überlegen. Es ist schade, daß nicht auch Herr MITCHELL Einiges dazu sagen kann. Die Tiefe der „osmotischen Schranke" innerhalb der Membran wird in der Regel zwischen 50 und 100 Å angenommen und wäre demnach von ähnlicher Größenordnung wie ein Enzym-Molekül. Aber auch eine wesentlich dickere Membran würde nicht ausschließen, daß der eigentliche Konzentrationsvorgang im Bereich des Enzyms erfolgt, denn der

Konzentrationsgradient braucht sich nicht über den gesamten Membranbereich zu erstrecken. Es kommt dabei auf die passive Impermeabilität der Schranke bzw. Transportregion an, in der das Enzym lokalisiert ist. Es wird ferner angenommen, daß die im Modell angegebenen Öffnungen neben dem Enzym zu eng sind, um freie Diffusion des Substrats durch die Schranke hindurch zuzulassen. Voraussetzung für den Durchtritt des Substrats ist eine spezifische Affinität zum Enzym und zum Acceptor. Die Affinität der Reaktion zwischen Substrat und Acceptor würde die Kraft liefern, die das Substrat durch die Schranke hindurchtreibt.

PAPE (Aachen): Ich möchte über einige klinische Versuche von NIEPER und BLUMBERGER und von uns mit einem 50:50-Gemisch von K^+- und Mg^{++}-Asparaginat berichten. Die unter Glykosidbehandlung beobachtete starke Erhöhung der Milchsäure und die wesentliche Erniedrigung der α-KGS wurden nach Verabfolgung von K-Mg-Asparaginat weitgehend ausgeglichen. Die BTS-Werte wurden durch Glykoside nicht beeinflußt. Extrasystolen als Ausdruck einer Intoleranz gegen Digitalis verschwanden nach Applikation von K-Mg-Asparaginat. Die Wirkung von K-Mg-Asparaginat wird als Verbesserung des oxydativen Potentials gedeutet. Nun ging aus Ihrem Vortrag hervor, daß die Asparaginsäure nicht aktiv in die Zelle einzudringen vermag?

HEINZ: Sowohl Asparaginsäure wie Asparagin werden aktiv transportiert, jedoch durch verschiedene Transportsysteme.

PAPE: Welches der gestern gezeigten Carrier-Modelle ist nach Ihrer Vorstellung geeignet, die geschilderte Wirkung der K-Mg-Asparaginsäure-Verbindung theoretisch zu erklären?

HEINZ: Man müßte zunächst klären, ob es sich dabei um einen chemisch-spezifischen oder nur um einen elektrostatischen Effekt handelt.

PAPE: Es handelt sich sicher um einen spezifischen Effekt, denn weder die Versuche mit Asparaginsäure noch die mit K-Asparaginat oder Mg-Asparaginat hatten die gleiche Wirkung wie die Kombination K-Mg-Asparaginat.

HEINZ: Nun ist das Asparaginat ein Anion, das beim Transport in die Zelle wahrscheinlich die Bewegung der Kationen elektrostatisch beeinflußt. Läßt es sich für Ihren Fall vollkommen ausschließen, daß der Eintritt des Kaliums in die Zelle lediglich durch einen derartigen elektrostatischen Effekt bewirkt wird?

PAPE: Wir bezeichnen diese Verbindung nicht als Carrier, sondern als „Elektrolytschlepper“. Wir wollen damit Elektrolyte in die Zelle hineinbringen und benutzen das Anion als „Schlepperverbindung“. Wir nehmen an, daß das K^+ durch eine Chelatbindung in die Zelle hineinkommt und dann — so hoffen wir — freigesetzt und als Ion wirksam wird.

HEINZ: Sie nehmen also eine Chelatbindung zwischen dem Asparaginat und den K-Ionen an, die in wäßriger Lösung nicht wahrscheinlich ist, aber für Lipoidphasen nicht von vornherein ausgeschlossen werden kann.

QUADBECK (Homburg): Läßt sich der Aminosäuretransport nicht auch durch eine Veresterung der Carboxylgruppe durch das Carriermolekül erklären? Hierfür würde u. a. die p_H-Abhängigkeit des Transports sprechen.

HEINZ: Grundsätzlich besteht eine solche Möglichkeit, vorausgesetzt, daß sie nicht mit einem Austausch von Sauerstoff verbunden ist.

KEPES (Paris): The p_H-dependance of the amino acid transport could indicate intervention of a histidine group. Dr. STEIN in DANIEL's Laboratory investigated the glycerol transport across the cell membrane and he found a p_H-dependance which also indicated histidine as the main side of linkage between the substrate and the transport mechanism. He evidenced this linkage by a 2-fluor-benzene reagent which reacts with the free histidine. This reaction is prevented when the transport substrate is present. I wonder, whether this experiment could not be done with your system also.

HEINZ: I thank you for the suggestion. The experiment should be done. We have, however, not been able to get a specific inhibition of glycine transport by diazotation, which also should attack imidazol groups.

WERLE (München): Sie hatten den Einfluß des Pyridoxals auf den Durchtritt der Aminosäuren durch die Zellmembran erwähnt. Man könnte folgenden Versuch anstellen: Es gibt Mikroorganismen, die in einem Vitamin B_6-freien Medium wachsen, aber bestimmte Leistungen nicht mehr vollführen können, zu denen sie nur imstande sind, wenn eben Pyridoxal in Form des Pyridoxalphosphates an bestimmte Fermente gebunden ist. Gibt es Hinweise dafür, daß in solchen Mikroorganismen der Influx oder der Efflux anders sind als in dem auf einem ausreichenden Nährboden gewachsenen Mikroorganismus? Viele Bakterien können z. B. eine Reihe von Aminosäuren außerordentlich schnell decarboxylieren. Bei Störung dieser Funktion könnte man sich vorstellen, daß etwa das Histidin oder Arginin, welches schwer decarboxyliert wird, mit verminderter Geschwindigkeit oder vielleicht überhaupt nicht durchtritt, weil eben die Voraussetzungen zur Einführung der Aminosäuren — nämlich die Überführung des Pyridoxals in Pyridoxalphosphat und die Bindung an ein Enzymprotein — fehlen. Dies wäre unabhängig davon, ob Pyridoxal vorhanden ist oder nicht.

HEINZ: Kürzlich hat HOLDEN in einer Veröffentlichung zu dieser Frage Stellung genommen. Bei einem bestimmten Mikroorganismus — ich glaube, Lactobacillus arabinosus — war die Anreicherung von Glutaminsäure bei spezifischem Vitamin B_6-Mangel stark beeinträchtigt. Da aber hierbei nicht die Anfangsgeschwindigkeit der Aufnahme, sondern lediglich der endgültige Verteilungsquotient des Glutaminats vermindert war, schien der B_6-Mangel nicht unmittelbar das Transportsystem zu betreffen. In einer späteren Mitteilung zeigte HOLDEN, daß bei Vitamin B_6-Mangel die Zellmembran geschädigt ist und daß diese Schädigung u. a. durch einen erhöhten osmotischen Druck im Außenmedium ausgeglichen werden kann. HOLDEN schließt daraus, daß das Pyridoxal nicht auf das Transportsystem, sondern auf die Zell-*Wand* einwirkt, und daß die Zellwand bei einer Vitamin B_6-Avitaminose schadhaft gebildet wird. Wegen dieser Schädigung, für die im Elektronenmikroskop auch morphologische Anhaltspunkte gefunden wurden, kann die Zellmembran anscheinend dem durch eine Anreicherung von Glutaminsäure hervorgerufenen osmotischen Druckgefälle nicht standhalten. Ich glaube jedoch nicht, daß diese Interpretation auch für die Ehrlich-Ascites-Tumorzellen zutrifft.

WERLE: Ist das Pyridoxal hier nur als eine Modellsubstanz aufzufassen oder als eine Substanz, die in physiologischer Weise bei dem Transport der Aminosäuren in die Zelle eine Rolle spielt?

HEINZ: Diese Frage läßt sich noch nicht befriedigend beantworten. Näheres darüber könnte man vielleicht mit Bakterien-Mutanten, bei denen die Umwandlung von Vitamin B_6 in Pyridoxalphosphat an irgendeiner Stelle gestört ist, erfahren. Soweit mir bekannt ist, sind solche Untersuchungen noch nicht gemacht worden.

RUMMEL (Homburg): Bei diesen Transportvorgängen interessiert naturgemäß die Art der Bindung des Substrats an den in Frage stehenden Carrier. Es scheint mir in diesem Zusammenhang interessant, den Aminosäuretransport dem Glucosetransport gegenüberzustellen. Der Glucosetransport ist obligat stereochemisch-spezifisch, was wir vom Aminosäuretransport nicht sagen können. Ich weiß nicht, ob hierin nicht ein Hinweis dafür liegt, daß die Art der Aminosäurebindung an den in Frage stehenden Carrier eine ganz andersartige ist.

HEINZ: Im Augenblick läßt sich nicht viel dazu sagen. Der Unterschied in der sterischen Spezifität der beiden Transportsysteme fällt zweifellos auf. Doch ist auch beim Aminosäuretransport die Spezifität scharf begrenzt. Wie wir bei den Versuchen mit Analogen des Glykokolls gefunden haben, bleibt die Transport-Affinität bei Einführung bestimmter chemischer Gruppen in das Molekül entweder voll erhalten oder sie schwindet vollständig. Wir finden kaum irgendwelche Übergänge. Die Spezifität ist zwar recht breit, aber doch ziemlich streng definiert.

DECKER (Hannover): Wir haben einen Transport von Harnstoff durch die Pansenwand bei Wiederkäuern nachgewiesen. Da die Transportrate bei einer Erhöhung des Bluthamstoffspiegels unverändert bleibt, dürfte es sich um einen aktiven Vorgang handeln. Sind bei Harnstoff, der hier nur wenig zur Sprache gekommen ist, Beobachtungen bekannt, die einen Hinweis auf die Natur eines eventuellen Carriermechanismus zulassen?

ULLRICH (Göttingen): Ein aktiver Harnstofftransport scheint in der Froschniere und in Nieren von Elasmobranchien (z. B. Haien) vorzuliegen. In der Niere von Warmblütern ist ein aktiver Harnstofftransport nicht sicher nachgewiesen worden.

NETTER: So, ich glaube, daß wir damit am Ende sind. Wir wollen nun das mehr oder weniger spezielle Kapitel, das natürlich auch allgemeine Ausblicke gestattet, verlassen und uns zum Ausklang noch einmal auf jene Mutterwissenschaft besinnen, von der aus eigentlich die gesamten osmotischen Erscheinungen, die Durchtrittserscheinungen, erforscht wurden. Von alters her ist da ja die scientia amabilis, und ich bitte unseren verehrten Vertreter dieser Wissenschaft, Herrn Prof. MOTHES, nun noch zu unserem Thema zu sprechen in der Formulierung seines Vortrags „Aktiver Transport als regulatives Prinzip für gerichtete Stoffverteilung in höheren Pflanzen“.

Aktiver Transport als regulatives Prinzip für gerichtete Stoffverteilung in höheren Pflanzen

Von

Kurt Mothes

Direktor der Botanischen Anstalten der Universität Halle/Saale

Mit 6 Textabbildungen

Im Leben der höheren Landpflanze spielt die Wasserversorgung der aktiven Gewebe eine entscheidende Rolle. Die Transpiration gefährdet stetig eine optimale Hydratur der Zellen. Saugkraftgefälle von den Orten besonderen Wasserbedarfs zu denen der Wasseraufnahme sorgen für gerichteten Wassertransport. Die Saugkraft der Zellen wird durch zwei Größen bestimmt: Durch den osmotischen Wert des Zellinhaltes, insbesondere des Zellsaftes der großen zentralen Vacuole, und durch die Spannung der im wesentlichen aus Cellulose- und Pektinmolekülen aufgebauten Zellwand. Es ist wahrscheinlich, daß diese Spannung einer gewissen Regulation unterliegt, indem unter dem Einfluß von Auxinen (z. B. Indol-3-essigsäure) zugunsten einer Erhöhung der plastischen und irreversiblen Dehnbarkeit die elastische gemindert werden kann.

Im allgemeinen folgt der Wassereinstrom in die Zellen diesen einfachen osmotischen Gesetzen. Von besonderer Wichtigkeit ist diese Wasserversorgung für die wachsende Zelle. Eine irreversible Vergrößerung der Zelle setzt Turgordehnung der Cellulose-Wand voraus. In der Organisation der höheren Pflanze finden wir deshalb ganz allgemein einen Saugkraftanstieg und einen Anstieg der osmotischen Werte zu den wachsenden Geweben hin. Andererseits darf als bewiesen gelten, daß von zwei Pflanzen gleicher Art diejenige besser wächst, die den geringsten osmotischen Wert und damit die höchste Hydratur besitzt. Offenbar ist die Pflanze gezwungen, die Sicherung der Wasserversorgung für wachsende „junge“ Gewebe mit einer Depression der Wuchsleistung zu erkaufen.

Aktive Zellen, das sind nicht nur meristematische, sondern auch speichernde Zellen, bedürfen aber auch der Zuführung einfacher Bau- und Betriebsmaterialien. Wasser und gelöste Substanz können sich an denselben Ort nicht gleichzeitig nach osmotischen Prinzipien bewegen. Schon daraus ist zu folgern, daß, wenn die eine Komponente osmotisch transportiert wird, die andere gegen einen Gradienten, also wahrscheinlich aktiv bewegt wird. Ein „aktiver" Transport des Wassers in Pflanzen ist oft behauptet, aber — wie mir scheint — niemals wirklich überzeugend bewiesen worden. Für einen „aktiven" Transport gelöster Substanzen gibt es aber eine große Zahl von Hinweisen.

So schreibt BRASSE schon 1886 über die Wanderung der Kohlenhydrate in der Zuckerrübe: In den Blättern waren 5% Zucker, in der Rübe aber 23—24%. Es gab keine befriedigende Erklärung für eine solche Wanderung von einer niederen zu einer höheren Konzentration: «... et je me crois autorisé à conclure de ce fait et de tout ce qui à servi à l'établir: Que le protoplasma de la racine peut contracter avec la saccharose une combinaison insoluble ou colloidale, mais en tout cas non dialysable et possédant une tension de dissociation. Que cette combinaison ne parait exister pendant la vie du végétal, et que, par conséquent, je ne crois pas qu'l soit possible de l'extraire pour le mieux définir, car le chloroforme qui tue ou anesthésie le protoplasma suffit pour détruire cette combinaison.»

LEONHARD (1939) weist darauf hin, daß selbst dann Zucker aus den Blättern abfließt, wenn er darin kaum als solcher nachzuweisen ist. PRINGSHEIM sagt schon 1906, daß es kaum eine Pflanze geben wird, bei der die jüngsten Teile im Zustande des Wassermangels nicht den ältesten Wasser zu entziehen vermöchten; die jüngeren Teile haben die höheren osmotischen Werte. Bei im Trockenen austreibenden Zwiebeln haben die austreibenden (also Material empfangenden) Teile höhere osmotische Werte als die Material abgebenden.

„Wäre es nur Rohrzucker selbst, so müßten wir einen Transport seiner Moleküle nach Stellen höherer Konzentration annehmen, was nur mit Energieaufwand unter aktiver Betätigung des Protoplasmas möglich wäre. Zu einer solchen Annahme liegt aber kein Grund vor." Er versucht dann eine Erklärung der Beobachtungen durch eine Umwandlung des auswandernden Rohrzuckers in Invertzucker und durch eine Bildung anderer osmotisch wirksamer Substanzen. Auch MASON u. PHILLIS (1934) versuchen die Schwierigkeiten einer Wanderung löslicher, N-haltiger Stoffe gegen ein Konzentrationsgefälle mit Stoffumwandlungen am Ziel der Translokation zu erklären. Auch wenn der wandernde Rohrzucker in Monosen, die wandernden Aminosäuren in Asparagin umgewandelt werden sollten, ist vielleicht die Tatsache, daß die Wanderung gegen das osmotische Gefälle bestehen bleibt, deshalb zu gering bewertet worden, weil von so bedeutenden Forschern wie URSPRUNG und BLUM (1916) auch einmal das Gegenteil behauptet worden ist: junge Blätter haben niedrigere osmotische Werte als ältere. Das ist aber wirklich eine Ausnahme (vgl. [5, 11, 12, 13, 34]). Auch bei WALTER (1931, 1950) finden sich zahlreiche Beispiele höherer osmotischer Werte in Geweben, die Ziel der Stoffwanderungen sind. Ganz besonders interessant sind jene Fälle eines Wechsels in der Hierarchie der osmotischen Werte. Wenn Rüben, Knollen usw. aufgefüllt werden, haben sie höhere osmotische Werte oder höhere Konzentrationen an Zucker oder Aminosäuren als die Blätter, wenn sie aber wieder austreiben, haben sie niedrigere Werte als die Triebe. [CURTIS 1935:...“the receiving tissues in all instances and by three methods of testing had higher osmotic concentrations than the supplying storage tissue.” (vgl. [33])].

LOOMIS (1945) betont, daß nicht der Bedarf an Stoffen (z. B. in einem verdunkelten, hungernden Blatt) die Stoffe dorthin lenkt, sondern im Gegenteil können solche Blätter noch den letzten Zucker abgeben. “A hypothesis of translocation in maize must not only account for movement against an osmotic gradient, but against gradients of each of the substances which might possible be translocated. Such secretory translocation certainly occurs between the leaf mesophyll and the phloem and probably along the phloem itself. Translocation in maize is polarized, out of the leaf, out of the xylem and toward the developing fruit. Polarized translocation out of the leaf is established during the later stages of tissue differentiation. Polarized translocation toward the fruit is established in the early phases of embryo development and does not develop in the absence of pollination.”

Auch dieser Stoffstrom in die Früchte und Samen ist nicht dirigiert durch den Verbrauch der löslichen Wanderstoffe zur

Synthese osmotisch unwirksamer, höher molekularer Verbindungen und durch ein auf solche Weise immer wieder hergestelltes Konzentrationsgefälle. Wenn man die Konzentrationen der Stoffe auf Wasser (und nicht — wie es meist geschieht — auf Trockengewicht) bezieht, so wird mit zunehmender Samenreife der Konzentrationsanstieg immer steiler. Einige schon 60 Jahre alte Belege aus einer Arbeit von EMMERLING mögen das für den löslichen N demonstrieren.

Tabelle 1. *Konzentration des löslichen Stickstoffs in % des Wassergehaltes* [nach Angaben von EMMERLING (1900) errechnet]

Datum	Wurzeln	Blätter	junge Blätter und Knospen	Hülsen	Samen
9. 6.	0,085	0,082	0,37		
12. 7.	0,119	0,144		0,22	0,46
10. 8.	etwa 0,110	etwa 0,07—0,22		0,15	0,35
30. 8.		0,11—0,18		0,14	0,68
10. 9.		0,22		0,55	1,1

FERNALD (1925) sieht bereits klar, daß diese Ordnung nicht nur im Zusammenhang steht mit der Wasserversorgung, sondern auch mit der eigentümlichen Hemmung des Wachstums seitlicher Knospen durch die Endknospe. "The data on the whole suggest a fairly close correlation between the osmotic concentration of a tissue and its tendency to inhibit growth of other tissues or to be inhibited by them." "Therefore the presence of a larger amount of osmotically active substances in a portion of a plant at the critical period would cause a diffusion of water into those tissues. Those portions of the stem may also have an advantage in obtaining the reserve foods because of their higher rate of metabolism."

Wenn man nun noch darauf hinweist, daß der Transport organischer Stoffe und der Prozeß der Auffüllung der empfangenden Gewebe abhängig ist von einer ergiebigen Atmung, daß er gehemmt wird durch HCN, DPN, so dürfte wahrscheinlich sein, daß es sich bei diesem Transport gegen ein Konzentrationsgefälle wirklich um das handelt, was man „aktiven Transport" nennt.

Die höhere Pflanze ist nun ein schlechtes Objekt zum Studium der Kinetik dieser Vorgänge, da der Protoplast der Pflanzenzelle gewissermaßen ein doppeltes Milieu hat: Die von außen herantretende Lösung und den Zellsaft der Vacuole. Auch wird eine summarische Gewebsanalyse nichts sagen über die wirkliche Konzentration eines Stoffes an einem Akkumulationsort. Als solche Orte kommen nicht nur die Vacuole selbst, sondern auch der Protoplast als ein Ganzes, osmotisch abgeschirmte Organelle wie Mitochondrien und Chloroplasten bzw. überhaupt Plastiden in Frage.

Um so mehr dürfte aber die höhere Pflanze ein überzeugendes Beispiel dafür abgeben, wie dieser „aktive Transport" den gesamten Zellstoffwechsel beherrscht, die Stoffaufnahme, die gerichtete Stoffwanderung, die Stoffspeicherung, das Wachstum und gewisse Korrelationen der Organe untereinander.

Um das richtig abzuschätzen, ist es nötig, auf die allgemeine Organisation der höheren Pflanze hinzuweisen. Sie stellt ein offenes System mit prinzipiell unbegrenztem Wachstum dar. Sie stirbt in ihren Teilen, aber Organtod bedeutet nicht Organismustod. Sie kann aus den noch überlebenden Teilen verlorengegangene ersetzen. Jedes wachsende Organ empfängt Bau- und Betriebsstoff von anderen, jedes ausgewachsene kann wie ein Speicher Stoffe abgeben bis zur Auszehrung. Alte, untere Blätter der Tabakpflanze geben Stoffe an die jüngeren, oberen Teile ab; das kann bis zur tödlichen Verarmung führen. Äußerlich ist dieser Prozeß am Vergilben der Blätter zu erkennen. Man nennt das auch „Altern". Dies beruht in erster Linie auf Überwiegen der Proteolyse und einer Abwanderung der Aminosäuren.

Dieses Vergilben ist ein durch Konkurrenz stark beeinflußbarer Vorgang. Entfernt man die jüngeren, vitaleren, attraktiveren Organe, bleiben die älteren Blätter länger grün. Befanden sie sich schon im Zustand des Vergilbens, kann man Wiederergrünen herbeiführen, indem man z. B. bei einer Tabakpflanze alle Teile oberhalb des vergilbenden Blattes abschneidet[23]. Man erreicht ähnliches, wenn man einem vergilbenden Blatt eine zusätzliche Stickstoffquelle eröffnet, z. B. indem man es über die Spreite, also unter Umgehung der Wurzel, direkt mit Stickstoffverbindungen versorgt. Daraus geht schon deutlich hervor, daß offenbar Konkurrenz um die Stickstoffverbindungen eine ausschlaggebende Rolle spielt[22].

Wenn man also alternde Blätter „verjüngen" kann, ist noch nicht geklärt, warum sich an der Pflanze das verschiedene Lebensalter der Organe in einer unterschiedlichen Konkurrenzfähigkeit auswirkt.

Auch wenn man darauf hinweist, daß das Wurzelsystem für die Erhaltung des jugendlichen Zustandes eines Blattes wesentlich ist, ist noch nichts darüber gesagt, warum an einer intakten, bewurzelten Pflanze die jungen Blätter vitaler sind als die älteren, die Früchte vitaler als die Blätter, der Stamm der herbstlichen Bäume vitaler als die Blätter usw.

Abb. 1. Isoliertes Blatt von Raps *(Brassica napus* L.*)*. Rechte Spreitenhälfte mit Kinetin besprüht. Nach 7 Tagen ist die linke Hälfte vergilbt, die rechte noch dunkelgrün und etwas gewachsen (dadurch nach links gebogen)

Schon lange bekannte Tatsachen deuten darauf hin, daß Hormone für diese eigentümliche Zuordnung der Teile und die damit zusammenhängenden, gerichteten Stoffströme von Bedeutung sind. Mit der Entdeckung des Kinetins wurde ein Modell für ein solches Hormon bekannt, das Stoffströme zu dirigieren vermag. Dieses 6-Aminofurfuryl-purin kommt vielleicht in Pflanzen nicht vor. Es wurde bei der Hydrolyse von Desoxyribonucleinsäuren erhalten und fand wegen seiner die Zellteilung fördernden Wirkung gegenüber bestimmten Gewebekulturen Beachtung. Es hat auch noch andere Wirkungen. Die allgemeinste und vielleicht bedeutsamste ist eben die der Förderung jenes Mechanismus, der für die Akkumulation und Attraktion von Stoffen verantwortlich ist.

Ein isoliertes, unbewurzeltes Blatt vergilbt auch unter günstigen Lichtbedingungen sehr schnell, obwohl keine jüngeren Organe als Konkurrenten vorhanden sein (Abb. 2). Der Stiel und die grobe Nervatur des Blattes wirken jetzt wie eine Konkurrent. Sie entziehen dem Parenchym der Blattspreite die Eiweißbausteine. Die Resynthese der Eiweiße ist deshalb nicht möglich. Besprüht man ein ganzes isoliertes, stielloses Blatt gleichmäßig mit Kinetin, so bleibt das Blatt grün. Gleichzeitig nimmt die Gesamtkonzentration an Aminosäuren und Amiden ab: Eiweiß wird synthetisiert (Abb. 2)[30, 24, 22, 7].

Besprüht man nur einen Teil eines solchen Blattes mit Kinetin, bleibt nur dieser Teil grün (Abb. 1,2). Aminosäuren-Chromatogramme zeigen aber außerdem, daß eine Akkumulation von Aminosäuren und Amiden stattfindet (Abb. 2 unten rechts). Diese löslichen Verbindungen entstammen den anderen Spreitenteilen.

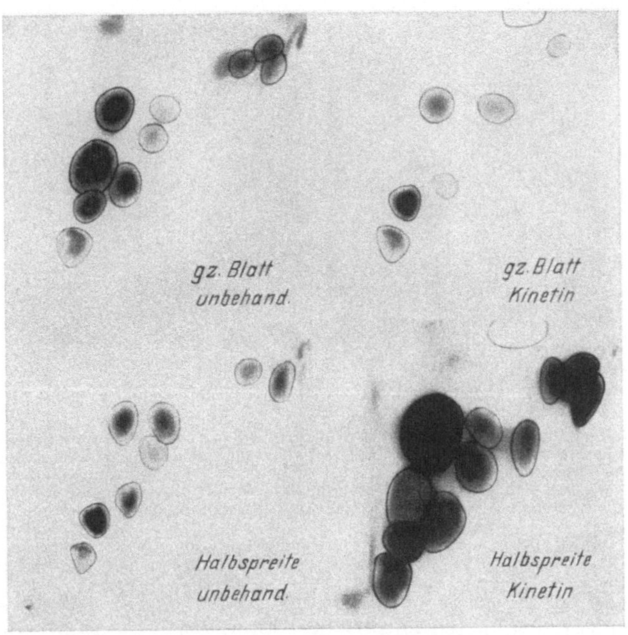

Abb. 2. Isolierte Blätter von Nicotiana rustica. 10 Tage nach Isolation und Lagerung bei schwachem Tageslicht. Aminosäuren-Chromatogramme (bezogen auf gleiches Frischgewicht). Laufmittel: von links nach rechts zweimal Propanol-Wasser, 3:1, nach oben: Phenol-Wasser, 4:1. Links oben: vergilbendes unbehandeltes Blatt. Rechts oben: gleiches Blatt, aber Spreite mit Kinetin besprüht, grün bleibend. Unten: Blatt auf linker Spreitenhälfte mit Wasser besprüht (vergilbt), auf rechter Spreitenhälfte mit Kinetinlösung, 30 mg/l (bleibt grün). Linke Hälfte gibt Aminosäuren ab, rechte akkumuliert sie (aus ENGELBRECHT 1961)

Jetzt konkurrieren also um die Eiweißbausteine der Stiel und das „Kinetin"-Gewebe. Gibt man genug Kinetin, gewinnt der Kinetinort. Der Kinetinort benimmt sich gegenüber dem isolierten, unbewurzelten und deshalb zum „Altern" verurteilten Organ wie ein junges Organ: er zehrt das alte aus. Die Wanderung der Aminosäuren gegen ein Konzentrationsgefälle kann besonders gut demonstriert werden durch die Applikation radioaktiv markierter Aminosäuren. Sie wandern schnell zum Kinetinort[24] (Abb. 3b).

Sie wandern auch dann dorthin und werden akkumuliert, wenn sie körperfremde Aminosäuren sind, die nicht in Eiweiß eingebaut werden (z. B. α-Amino-iso-buttersäure, D-Leucin).

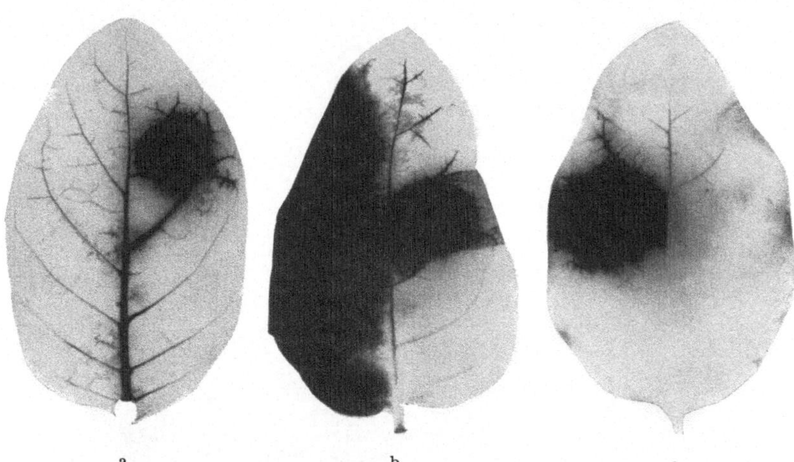

Abb. 3. Radioautogramme isolierter Blätter von Nicotiana rustica. a) Rechte Spreitenhälfte erhält auf kreisförmig begrenztem Bezirk ^{14}C-Glycin appliziert. b) Wie a, aber außerdem erhält die linke Spreitenhälfte Kinetin. Die „Radioaktivität" wandert zum Kinetinort. c) Die linke Hälfte erhält Kinetin, aber außerdem auf kreisförmig begrenztem Bezirk ^{14}C-Glycin. Die „Radioaktivität" breitet sich nicht aus (Retention der Aminosäuren)

Nicht eine Massensynthese von Eiweißen zieht die Aminosäuren nach, sondern der aktive Prozeß einer Akkumulation löst ein Nachströmen von Aminosäuren aus und kann auch — wenn alle notwendigen Bedingungen günstig sind — zur beachtlichen Eiweiß-(Netto)-Synthese führen [25, 27].

Kinetingewebe verhält sich noch in anderer Beziehung wie junges Gewebe. Es attrahiert nicht allein verschiedenste lösliche Stoffe, es hält sie auch fest. Deshalb bleibt der hohe osmotische Wert des jugendlichen Gewebes im Organverband erhalten, und deshalb regenerieren junge Blätter z. B. keine Wurzeln. Will man aus einem Blattstiele Wurzeln hervorlocken, braucht man Auxin und ein (z. B. durch Hungern) „gealtertes", d. h. geschwächtes, Blatt, dessen Retentionsvermögen gering ist. Dieses gibt nun u. a.

Aminosäuren ab, die basalwärts wandern und das Wurzelwachstum gestatten. Besprüht man ein Blatt mit Kinetin, so werden diese Stoffe festgehalten, die Wurzelbildung ist erschwert oder völlig unterbunden[22, 10].

Sprüht man auf ein in Wurzelbildung befindliches Blatt eine zur Eiweißsynthese unbrauchbare, radioaktiv markierte Aminosäure (z. B. α-Aminoisobuttersäure), so taucht die Radioaktivität schnell in den Wurzelspitzen auf. Sie zeigt an, daß auch andere Aminosäuren dorthin wandern.

Sprüht man gleichzeitig Kinetin auf die Blattfläche, so ist die Abwanderung zur Wurzelspitze erschwert. Die Wurzeln müssen unter Umständen aus Mangel an Baumaterial das Wachstum einstellen[10].

LETTRÉ (1958) sagt, daß Zellen vom Yoshida-Tumor der Ratte in Gewebekultur nur begrenzt lebensfähig sind. Gibt man aber Fibroblastenkultur hinzu, so wandern die Tumorzellen zu den Fibroblasten, siedeln sich an und vermehren sich. Wahrscheinlich erhalten sie von diesen ,,Wirten" besondere Stoffe. Setzt man einer solchen Mischkultur das kinetinanaloge 6-Purylhistamin zu, so wird das Wachstum der Tumorzellen gehemmt, sie gehen zugrunde. Die normalen Bindegewebszellen überleben. Es liegt nahe, die Zellen der Wurzelspitzen eines regenerierenden isolierten Blattes mit den Tumorzellen und das Blattgewebe mit der Fibroblasten-Kultur zu vergleichen. Nur wenn die attraktiven und zurückhaltenden Kräfte in einem richtigen Verhältnis stehen, wird sich ein Stoffstrom von einem zum anderen Orte auslösen lassen.

Wir demonstrieren diese Fähigkeit der Kinetin-Gewebe, Stoffe anzuziehen, zu akkumulieren und festzuhalten, noch durch einen Versuch mit isolierten Blättern von Nicotiana rustica. Es ist schon verschiedentlich gezeigt worden, daß in einem unbewurzelten Blatt radioaktiv markierte Verbindungen nur sehr schwer von einer Hälfte in die andere wandern[1, 15, 21, 29]. Applizieren wir der rechten Spreitenhälfte eines isolierten Blattes von Nicotiana rustica eine ^{14}C-markierte Aminosäure auf beschränkter, kreisförmig begrenzter Fläche, so wandert die Radioaktivität vor allem in den Nerven basalwärts (Abb. 3a). Wird der linken Spreitenhälfte Kinetin appliziert, so wandert schnell ein erheblicher Teil der ,,Radioaktivität" in den Kinetinbezirk. Die sonst bestehende Wanderungsschwierigkeit ist also durch Schaffung eines Attraktionszentrums mit Hilfe von Kinetin überwunden (Abb. 3b). Wird einem gleichen Blatt die ^{14}C-markierte Aminosäure jedoch links appliziert, d. h. in die Kinetin-Spreitenhälfte, dann breitet sich die Radioaktivität nicht nach rechts aus und nicht einmal in die anderen Bezirke der Kinetinhälfte (Retention der Aminosäure durch „jugendliche" „aktive" Organe) (Abb. 3c).

Ich meine, daß dieses Beispiel in geradezu schematischer Weise die verschiedenen Möglichkeiten stofflicher Beziehungen zwischen den Organen darlegt und — ohne Gewalt anzutun — auf das hier erörterte Problem der Beziehungen zwischen jungen und alten Teilen, zwischen Früchten, heranwachsenden Speichern und assimilierenden Teilen, sich entleerenden Speichern und austreibenden Teilen bezogen werden kann.

Abb. 4. Nicotiana rustica-Pflanze. Blatt (rechts) 14 Tage lang verdunkelt, es vergilbt. Jedoch bleibt die rechte Spreitenhälfte, die mit Kinetin behandelt ist, trotz Verdunkelung grün

Es ist leider völlig unbekannt, wie das Kinetin den Akkumulationsmechanismus auslöst. Folgende Hinweise können vielleicht zu einer Deutung des Gesamtphänomens beitragen:

Werden isolierte Blätter partiell verdunkelt, findet eine Wanderung von Aminosäuren zu den Licht-Orten statt. Ein belichteter Blatteil verhält sich also wie ein Kinetinbezirk[7]. Wenn an der Pflanze befindliche Blätter gut belichtet sind, zeigt Kinetin kaum sichtbare Wirkungen. Wenn an einer normal belichteten Pflanze ein einzelnes Blatt weitgehend verdunkelt wird, so vergilbt es in einigen Tagen. Wird dieses Blatt über die ganze Spreite oder über einen Teil von ihr mit Kinetinlösung besprüht, so bleibt dieser Kinetinbezirk auch im Dunkeln lange Zeit (länger als einen Monat) grün (Abb. 4). Er ernährt sich durch Attraktion von Stoffen aus den belichteten Blättern. Bei isolierten, belichteten Blattstückchen stellt sich die akkumulative Wirkung des Kinetins nicht sofort

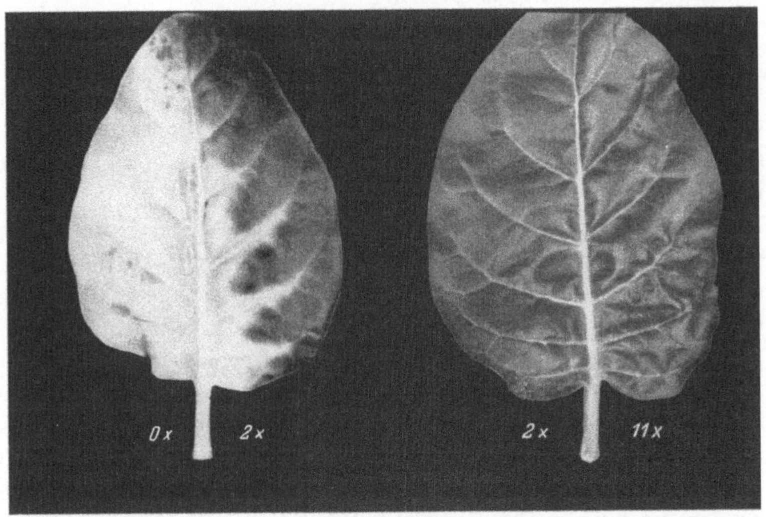

Abb. 5. Isolierte Blätter von Nicotiana rustica 13 Tage nach der Isolation. Links: rechte Spreitenhälfte erhielt zweimal Kinetin, linke Wasser: diese vergilbt. Rechts: Beide Hälften erhalten Kinetin, aber verschieden oft (zweimal bzw. elfmal) (MOTHES 1960)

ein. Man gewinnt den Eindruck, die Blätter müssen erst an einem natürlichen, dem Kinetin ähnlichen Faktor verarmen. Kinetin wirkt aber auch bei verdunkelten, isolierten Blättern[26].

Offenbar bildet sich bei grünen, intakten Blättern im Licht auch ohne Kinetin das akkumulierende System aus. Ob diese Lichtwirkung über die Synthese einer in der Wirkung dem Kinetin ähnlichen Substanz geht oder direkt (z. B. durch Photo-Phosphorylierung über die Bildung von ATP als Voraussetzung der Akkumulation), ist noch nicht geklärt. Jedenfalls gibt es so etwas wie eine maximale Aktivierung, eine „Sättigung" dieses akkumulativen Systems. Eine solche „Sättigung" kann man z. B. auf folgende Weise demonstrieren (Abb. 5). Besprüht man die rechte Hälfte eines isolierten Blattes zweimal mit Kinetinlösung (20 mg/l)die linke mit Wasser, so bleibt — wie schon gezeigt — die rechte grün, die linke wird ausgezehrt und vergilbt. Besprüht man eine Hälfte zweimal mit Kinetin, die andere elfmal, so verhalten sich beide Hälften gleich. Obwohl die eine Hälfte also mehr als fünfmal so viel Kinetin erhalten hat, ist sie nicht überlegen. Eine maximale Aktivierung des akkumulativen Systems wird also bereits mit zweimaliger Kinetingabe erreicht[22].

Diese Beobachtungen lenken auf eine andere, wie mir scheint, wichtige Beobachtung hin. Es gibt vielerlei an sich bescheidene und keineswegs tödliche Noxen, die mehr eine Schwächung als eine Schädigung hervorrufen. Aber im Verbande kann ein geschwächtes Organ so in Nachteil geraten, daß es bei der Konkurrenz um die Stoffe unterliegt.

Wenn man eine Hälfte eines Tabakblattes eine Minute lang auf 50° C erhitzt, so bleibt diese Hälfte turgescent. Sie hat nicht sichtbar gelitten, aber vergilbt ungewöhnlich schnell. Die nicht erhitzte Hälfte verhält sich wie eine mit Kinetin versorgte Hälfte. Sie ist in bezug auf den Akkumulationsmechanismus überlegen und entreißt der durch Hitze geschwächten Hälfte u. a. Aminosäuren. Appliziert man einer Spreitenhälfte eines jungen Blattes auf begrenzter, kreisförmiger Fläche ^{14}C-markiertes Glycin, so breitet es sich kaum aus, es wird festgehalten. Das akkumulative System ist noch genügend intakt. Wird dieses Blatt auf subletale Temperatur erhitzt, so bleibt es zunächst äußerlich ganz gesund. Das applizierte Glycin wandert aber zur anderen, noch vitaleren Seite.

Das akkumulative System ist also sehr empfindlich. Seine Schwächung bedeutet nicht nur mangelnde Fähigkeit zur Attraktion, sondern auch mangelnde Fähigkeit zur Stoffretention.

Wenn die Beeinträchtigung dieses Systems im Rahmen einer reversiblen Schwächung bleibt, dann dürfte erwartet werden, daß Kinetin einer solchen Schwächung entgegenwirkt. Das ist in geradezu eklatanter Weise der Fall. Es ist dabei gleichgültig, ob Kinetin vor oder nach dem Erhitzen appliziert wird.

In Abb. 6 ist die rechte Blatthälfte erhitzt. Dieselbe Hälfte erhält im oberen Viertel Kinetin. Dieses Viertel bleibt nicht nur länger grün als der rechte untere Quadrant, es bleibt auch dunkler als die nicht erhitzte linke Hälfte. Die Verteilung der Stickstoffverbindungen (Chromatogramme; ^{15}N-Versuche) zeigen ganz klar, daß dieser Kinetin-Quadrant trotz Erhitzung vitaler ist als alle anderen Blatteile. Gibt man einem solchen Blatt über den Stiel ^{14}C-markierte Aminosäuren, so werden diese in dem Kinetin-Quadranten akkumuliert. Kinetin wirkt also wie ein Faktor der Hitzeresistenz[9].

Ähnliches scheint sich auch beim Welken der Blätter abzuspielen. Wir haben früher den raschen Eiweißzerfall in welkenden

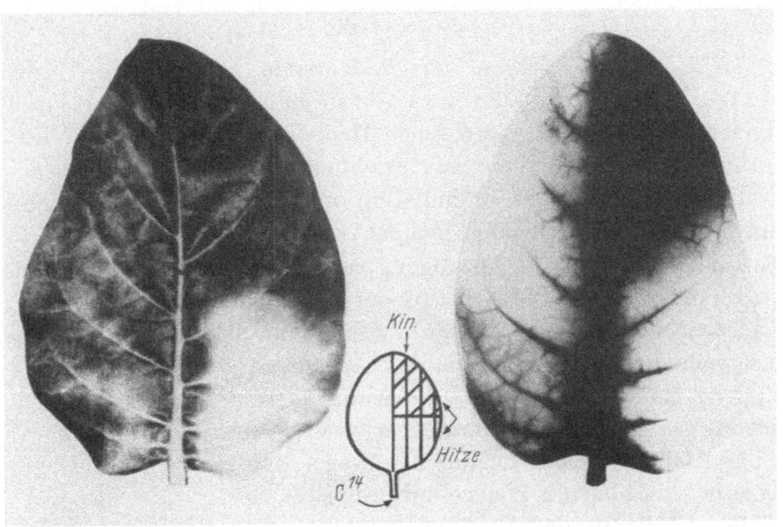

Abb. 6. Isoliertes Blatt von Nicotiana rustica. Rechte Seite für 60 sec auf 50° C erhitzt. Rechts oben Kinetin appliziert (dieser Quandrant bleibt am längsten grün). Über den Stiel wird gleichzeitig ^{14}C-α-Aminoisobuttersäure appliziert. Diese akkumuliert vor allem in Kinetin-Quadranten rechts oben. (Links: Photographie; rechts: Radioautographie) (ENGELBRECHT u. MOTHES 1960)

Blättern beschrieben und den Abtransport der löslichen Spaltprodukte in junge noch turgescente Pflanzenteile. Wir standen damals noch unter dem Einfluß der Vorstellung, daß Abbau und Synthese der Eiweiße durch das gleiche Enzymsystem bedingt sind, daß aber Synthese in höherem Maße strukturgebunden ist als der Abbau. Beim Welken soll diese Struktur gestört werden. Diese Vorstellungen haben einige Wandlungen erfahren. Und mir scheint, daß jetzt ein weiteres wichtiges Moment beigesteuert werden kann. Welken stört das empfindliche akkumulative System.

Wenn die stetig anfallenden Eiweißspaltprodukte aber nicht zurückgehalten werden können, wird auch die Eiweißresynthese nicht möglich sein. Eiweißabbau überwiegt.

Erst die Inkorporationstechnik radioaktiv markierter Aminosäuren gestattet uns einen tieferen Einblick in die wirklichen Geschwindigkeiten der Synthese. Hierzu werden sich in Kürze meine Mitarbeiter PARTHIER und WOLLGIEHN äußern. Sie können zeigen, daß die Synthese von Eiweiß und Nucleinsäure wesentlich durch das Ausmaß der Stoffakkumulation mit bestimmt wird. Kinetin fördert primär diese Akkumulation und erst sekundär die Massensynthese.

1887 (S. 355) schrieb der bedeutende Pflanzenphysiologe JULIUS SACHS: „So können wir also mit schematischer Vereinfachung sagen: jeder wachsende Theil der Pflanze wirkt auf die vorhandenen Baustoffe wie ein Anziehungscentrum, jeder Reservestoffbehälter und jedes Assimilationsorgan dagegen verhält sich einem wachsenden Theile gegenüber passiv oder allenfalls wie ein Abstoßungscentrum. Die Anregung zu den Stoffbewegungen aber wird immer durch das Wachstum der jungen Organe gegeben; die Knospen eines Baumes treiben im Frühjahr nicht etwa deshalb aus, weil, wie die Leute sagen, der Nahrungsstoff in sie eindringt, sondern gerade umgekehrt: die Nahrungsstoffe werden in Bewegung gesetzt, weil die Knospen zu wachsen anfangen."

Das läßt uns zum Schluß die Frage nach dem „natürlichen Kinetin" der höheren Pflanze aufwerfen.

Es ist bekannt, daß Auxin (β-Indolylessigsäure) wie auch einige analoge Verbindungen ähnliche Wirkungen wie das Kinetin hervorrufen. Wir haben, soweit das Grünbleiben sonst vergilbender Blätter betroffen ist, darauf schon früher ausführlicher hingewiesen[22]. Ich möchte jetzt noch die Arbeit von YAKUSHKINA, POROISKAJA und FILATOWA (1956) erwähnen. Nachdem schon KURSANOW und SAPROMETOW (1949) gezeigt hatten, daß in abgeschnittenen Getreidehalmen Aminosäuren sich sehr schnell, schneller als der Transpirationsstrom, aufwärts gegen das Konzentrationsgefälle bewegen, konnten YAKUSHKINA u. Mitarb. mitteilen, daß diese Aufwärtsbewegung unterbleibt, wenn gleichzeitig Auxin der Stammbasis zugefügt wird. Dieses Auxin bleibt in den unteren Teilen liegen, Gleichzeitig findet im Auxingewebe eine Akkumulation löslicher Stoffe statt. Daß diese Auxinwirkungen im allgemeinen nicht so markant sind wie beim Kinetin, hängt wohl in erster Linie an zwei Vorzügen des Kinetins: es bleibt dort liegen, wo es appliziert wurde, und es wird nur langsam abgebaut.

Auxin wandert schnell ab und unterliegt schneller Umwandlung.

Fräulein Dr. ENGELBRECHT (1961), mit der ich den größten Teil der hier vorgetragenen Experimente durchgeführt habe, hat mit ^{14}C-markiertem Auxin beobachtet, daß die akkumulative Wirkung vielleicht nicht vom Auxin selbst, sondern von einem Derivat ausgeht, das einen R_f-Wert ähnlich der Indol-acetyl-asparaginsäure hat. Mit verschiedenen mobilen Phasen läßt sich zeigen, daß diese Fraktion aus einem Gemisch verschiedener peptidartiger Verbindungen der Indolylessigsäure mit Aminosäuren vor allem von α-Amino-dicarbon-säure-Typ besteht. Welche dieser Verbindungen kinetinanalog wirken, muß noch geprüft werden. Es spricht einiges dafür, daß die Beeinflussung der Plastizität der Zellwand und die des akkumulativen Systems durch zwei verschiedene, aber nahe verwandte Stoffe ausgeht. Das wird die Zukunft klären müssen. Wenn man nun noch auf die eben erschienene Arbeit von RIGGS und WALKER (1960) hinweist, die wahrscheinlich macht, daß das Wachstumshormon (das somatotrope Hormon) der Säugetiere die Akkumulation von α-Aminoisobuttersäure in verschiedenen tierischen Geweben fördert, und wenn man weiter beachtet[3], daß die Akkumulation von Glycin in Tumorzellen durch 10—20 mM Indolylessigsäure stark gefördert wird (gewiß unter pflanzenphysiologischen Gesichtspunkten eine sehr hohe Konzentration), so ergibt sich, daß die Wirksamkeit des aktiven Transportes durch recht verschiedene Stoffe wesentlich erhöht wird. Vielleicht darf man deshalb vermuten, daß alle diese „Kinine" eine sehr allgemeine Funktion treffen. Vielleicht ist ihre Wirkung mehr physikalisch-chemischer Art, wie sie von VELDSTRA für das Auxin postuliert wurde, und nicht spezifisch-chemischer Art.

Soviel also auch nebelhaft erscheint, ich meine, daß das Prinzip des aktiven Transports und damit das Prinzip gerichteten Stofftransports und betonter Stoffretention einer der entscheidenden Faktoren für das Verhalten der pflanzlichen Organe zueinander und für den gesamten Entwicklungsablauf einer höheren Pflanze überhaupt ist.

Literatur

[1] ARONOFF, S.: Plant Physiol. **30**, 184 (1955).
[2] BRASSE, L.: Ann. Agron. **12**, 305 (1886).
[3] CHRISTENSEN, H. N., TH. R. RIGGS, H. FISCHER and J. N. PALATINE: J. Biol. Chem. **198**, 17 (1952).
[4] CURTIS, O. F.: The translocation of solutes in plants. 273 S. New York 1935.
[5] CURTIS, O. F., and H. T. SCOFIELD: Amer. J. Bot. **20**, 502 (1933).
[6] EMMERLING, A.: Landw. Vers. Stat. **54**, 215 (1900).
[7] ENGELBRECHT, L.: Flora **150**, 73 (1961).
[8] ENGELBRECHT, L.: Ber. dtsch. bot. Ges. **74**, 1961 (im Druck).
[9] ENGELBRECHT, L., u. K. MOTHES: Ber. dtsch. bot. Ges. **73**, 246 (1960).
[10] ENGELBRECHT, L., u. K. MOTHES: Plant and Cell. Physiol 1961 (im Druck).
[11] FERNALD, E. I.: Amer. J. Bot. **12**, 287 (1925).
[12] HARRIS, J. A., R. A. GARTNER and J. V. LAWRENCE: Bull. Torrey Bot. Club. **44**, 267 (1917).
[13] HERRICK, E. M.: Amer. J. Bot. **20**, 18 (1933).
[14] KURAISHI, S.: Sci. Pap. Coll. Gen. Educ. Univ. Tokyo **9**, 67 (1959).
[15] KURSANOW, A.: Die Kulturpflanze. Beih. **1**, 101 (1956).
[16] KURSANOW, A., u. M. N. SAPROMETOW: Dokl. Akad. Nauk. SSSR **68**, 1113 (1949).
[17] LEONHARD, D. A.: Plant Physiol. **14**, 55 (1939).
[18] LETTRÉ, H.: Wien. Med. Wschr. **1958**, 791.
[19] LOOMIS, W. E.: Science **101**, 398 (1945).
[20] MASON, T. G., and E. PHILLIS: Ann. Bot. **48**, 315 (1934).
[21] MOTHES, K.: Die Kulturpflanze. Beih. **1**, 101 (1956).
[22] MOTHES, K.: Naturwiss. **47**, 337 (1960).
[23] MOTHES, K., u. W. BAUDISCH: Flora **146**, 521 (1958).
[24] MOTHES, K., u. L. ENGELBRECHT: Mber. dtsch. Akad. Wiss. **1**, 367 (1959).
[25] MOTHES, K., L. ENGELBRECHT and H. R. SCHÜTTE: Physiol. Plant. **14**, 72 (1961).
[26] MOTHES, K., u. L. ENGELBRECHT: Photychemistry **1**, 1961 (im Druck).
[27] PARTHIER, B.: Flora **151**, 1961 (im Druck).
[28] PRINGSHEIM, E.: Jb. wiss. Bot. **43**, 89 (1906).
[29] RABIDEAU, G. S., and G. O. BURR: Amer. J. Bot. **32**, 349 (1945).
[30] RICHMOND, A., and A. LANG: Science **125**, 650 (1957).
[31] RIGGS, TH. R., and. L. M. WALKER: J. Biol. Chem. **235**, 3603 (1960).
[32] SACHS, J.: Vorlesungen über Pflanzenphysiologie. 2. Aufl. (Leipzig 1887).
[33] STREET, H. E., A. E. KENYON and G. M. WATSON: Ann. Appl. Biol. **33**, 1 (1946).
[34] STOCKING, C. R.: Handbuch der Pflanzenphysiologie. Bd. 2, 57 (1956).
[35] URSPRUNG, A., u. G. BLUM: Ber. dtsch. bot. Ges. **34**, 88 (1916).
[36] VELSTRA, H.: Ann. Rev. Plant Physiol. **4**, 151 (1953).
[37] WALTER, H.: Die Hydratur der Pflanze. 174 S. Jena: Fischer 1931.
[38] WOLLGIEHN, R.: Flora **151**, 1961 (im Druck).
[39] YAKUSCHKINA, N. J., S. M. POROISKAJA i T. G. FILATOWA: Fis. Rast. **3**, 423 (1956).

Unsere eigenen Arbeiten haben eine zusammenfassende Darstellung erfahren [K. MOTHES, Naturwiss. **47**, 337 (1960)]. Wir haben hier nur neuere oder dort nicht genügend berücksichtigte Veröffentlichungen zitiert.

Diskussion

Diskussionsleiter: NETTER, *Kiel*

Lieber Herr MOTHES, ich danke Ihnen ganz besonders herzlich für diesen wunderschönen Abschlußvortrag, den Sie uns geboten haben. Sie haben uns nicht beraubt, sondern im Gegenteil, Sie haben uns beschenkt, und dafür danke ich Ihnen herzlich. Und Sie haben uns gezeigt, wie hier der aktive Transport als ein sehr umfassendes Prinzip in der Gesamtpflanze wirksam ist, und haben weiter auch Stoffe aufgezeigt, die als Generalschlüssel gewissermaßen hier in Erscheinung treten und von denen wir also zu diskutieren hätten, ob dieser Generalschlüssel doch irgendwie — und wie er es tut — in den aktiven Transport eingreift. Ich weiß, daß die Zeit sehr weit vorgeschritten ist, aber dennoch eröffne ich die Diskussion, denn hier sind so viele Fragen zu erwarten, daß mindestens mit dem Versuch einer gewissen Auslese bei den Fragen, die Sie selbst treffen mögen, doch das Fragen erlaubt ist. Ich darf Sie bitten.

HERRMANN (Berlin): Nach bisher unveröffentlichten Versuchen von Herrn Dipl.-Lebensmittelchemiker VOGEL an meinem Institut wird ^{32}P aus anorganischem Phosphat sehr schnell in den Keimen und direkt unter der Oberfläche von ganzen Kartoffeln angereichert. Die markierte Phosphatlösung wurde in ein mittels Korkbohrer geschaffenes etwa 1 cm tiefes Loch am Nabel der Kartoffel eingeschleust. Im Autodiagramm konnte man an anschließend hergestellten Kartoffelschnitten die obige Verteilung feststellen, während das Innere der Kartoffel wenig Aktivität aufwies. Die Untersuchungen fanden an z. Zt. keimenden Kartoffeln bei Zimmertemperatur statt. An der Bestimmung, wie weit das eingelagerte anorganische Phosphat als Zuckerphosphorsäureester, ATP usw. vorliegt, wird gegenwärtig noch gearbeitet. Auf Grund dieses Sachverhaltes möchte ich zwei Fragen an den Vortragenden stellen:

1. Wie kann man sich die schnelle Einlagerung des Phosphates unter der Scholle erklären?

Das Aktivwerden der Keime leuchtet ohne weiteres ein.

2. Aus dem Vortrag glaube ich entnommen zu haben, daß die Wanderung von Nährstoffen zu den Orten höheren osmotischen Druckes verlaufen soll. Wäre demnach anzunehmen, daß unterhalb der Schale die Zellen höhere osmotische Drucke als im Kartoffelinneren aufweisen?

Wir konnten, ohne jedoch damals eine so dünne Oberflächenschicht zu berücksichtigen, früher keine signifikanten Unterschiede zwischen den Gefrierpunktsdepressionen von Kartoffelstückchen finden, die an verschiedenen Punkten der Kartoffelknolle (Krone, Mitte, Nabel, Seite) entnommen waren.

MOTHES: Das ist untersucht. Die osmotischen Werte der Kartoffelknolle schwanken. Nach der Ruhe ist der Wert in den „Augen“ und den daraus sich entwickelnden Trieben höher als im Speicherparenchym. Auch da gibt es Unterschiede. Literatur findet man bei FERNALD, CURTIS, STREET. Ich bin geneigt zu sagen, daß im allgemeinen gilt: Wo die höheren osmotischen Werte sind, dorthin geht auch der gerichtete, aktive Transport gelöster Stoffe. Das Phosphat müßte also am stärksten in den Trieben akkumuliert sein.

MANDEL (Straßburg): Haben Sie denselben Effekt des Kinetins mit jungen Blättern?

MOTHES: Mit jungen Blättern erreicht man nicht ohne weiteres sichtbare Kinetin-Effekte. Wir nehmen an, daß sie einen kinetin-analogen Faktor in genügender Menge besitzen. Das ist eben das Merkmal ihrer Jugendlichkeit. Wenn junge Blätter einige Tage isoliert sind, verarmen sie an verschiedenen Stoffen. Dann reagieren sie allmählich auch gut auf Kinetin.

FISCHER (Frankfurt): Ich weiß nicht, ob Herr Professor MOTHES erlaubt, ein Bild zu zeigen, was nichts mit Kinetin zu tun hat, aber zeigt, daß man an pflanzlichen Zellen Transportvorgänge beeinflussen kann mit Substanzen, die auch an tierischen Zellen vielartige Transportsysteme beeinflussen, und zwar im hemmenden Sinne, Prednisolon und andere Steroide. Läßt man Bohnen in Anwesenheit von Prednisolon keimen, so sieht man, daß die Keimung praktisch unterbleibt. Das ist eine hohe Konzentration von Solodecortin, also Prednisolon. Setzt man sie herab, so beobachtet man, daß vor allem die Ausbildung der Wurzel in ihrer Verzweigung unterbleibt, während der nach oben gehende Stamm sich gleichmäßig, fast gleichermaßen wie die Kontrolle entwickelt. Wir hatten damals gedacht — das hat sich aber nicht bewahrheitet —, daß man in den Steroiden fast ein Prinzip, einen Gegenstoff zum Kinetin habe. Und auch jetzt versuchen wir noch die Beziehungen zu den Auxinen herzustellen. Es hat sich aber noch nichts Genaues herausgestellt dabei. Sicher ist, daß auch hier die Beeinflussung der aktiven Transportvorgänge vor sich geht.

MOTHES: Zunächst liegt uns selbst die Frage nahe, warum zwei chemisch so verschiedene Substanzen wie Auxin und Kinetin ähnliche Wirkungen haben. Es ist sowohl vom Kinetin als auch von der Indolylessigsäure eine große Zahl von Analogen synthetisiert worden. Wenn man beim Kinetin den Furfurylrest hydrophil macht oder den Purinrest lipophil (etwa durch einen Benzol-Rest ersetzt), so ist in beiden Fällen die Wirkung aufgehoben. Das gleiche gilt für das Auxin. Wird am Indol-Ring eine OH-Gruppe eingebaut oder umgekehrt der hydrophile Charakter der Seitenkette gelöscht, schwindet die Auxin-Wirkung. Es scheint also für beide Substanzen entscheidend zu sein, daß in einer bestimmten Anordnung eine lipophile und eine hydrophile Gruppierung kondensiert sind. Ich bin der Meinung, daß diese Stoffe nicht unmittelbar in den aktiven Transport eingreifen, daß sie also nicht selbst carrier sind, sondern daß sie entweder physikalisch-chemisch wirken und den Zustand der Membranen oder den Transport des ATP beeinflussen, ohne den ein aktiver Transport nicht möglich ist. Wichtig wäre zunächst, Klarheit darüber zu erlangen, wo der aktive Transport stattfindet und wo die Akkumulation erfolgt. Für die Mitochondrien ist das schon nachgewiesen. Für die Plastiden ist es sehr wahrscheinlich. Daß im Licht grüne Organe besonders leicht akkumulieren, könnte vielleicht eine Erklärung finden, wenn diese Akkumulation in Chloroplasten selbst erfolgen und durch die ATP-Bildung im Zuge der Photophosphorylierung gefördert würde. Wir sind bereits zum Studium dieser Erscheinung an isolierten Chloroplasten übergegangen.

Holzer (Freiburg): Ich wollte nur fragen, ob die Effekte auch an Mikroorganismen, Hefe, Schimmelpilzen und ähnlichen zu sehen sind?

Mothes: Mit Kinetin selbst ist bisher an Mikroorganismen wenig gearbeitet worden, wenigstens nicht im Sinne der Akkumulationseffekte. Gale, ein englischer Mikrobiologe, der den aktiven Transport bei Bakterien untersucht hat, bemerkt 1958, daß mit Ultraschall zerstörte Bakterien nach dem Waschen unfähig sind, Aminosäure aufzunehmen und zu inkorporieren. Wenn er aber einen „Inkorporations-Faktor" hinzugibt, den er durch enzymatischen Abbau von RNS erhält, so ist die Inkorporation wieder möglich. Ich könnte mir denken, daß es sich hier um den Effekt einer kinetin-analogen Substanz handelt. Wir selbst sind dabei, einen mikrobiellen Test für das Kinetin auszuarbeiten. Vielleicht werden Sie sagen, man hätte gleich mit Mikroorganismen arbeiten sollen, um den Mechanismus der Kinetinwirkung aufzuhellen. Ich muß aber freimütig bekennen, daß wir gar nicht die Absicht hatten, über den aktiven Transport zu arbeiten. Vielmehr wollten wir wissen, warum isolierte Blätter schnell altern und vergilben und warum sie es nicht tun, sobald sie Wurzeln haben. In solchem Zusammenhang benutzten wir auch Kinetin, das als ein Zellteilungshormon deklariert worden war. Es veranlaßte aber bei unseren Blättern keine Zellteilungen. Doch wirkte es wie ein Verjüngungsfaktor. Und das erschien uns auch interessant. Und so kamen wir zum aktiven Transport.

Netter: Ich glaube, es sind keine weiteren Wortmeldungen vorhanden. Damit komme ich jetzt dazu, die Tagung zu schließen. Nunmehr möchte ich allen, die sich an der Diskussion beteiligt haben, vor allem aber allen Vortragenden, besonders denen, die eine sehr weite Reise zum Vortragsort gehabt haben, noch einmal recht herzlich dafür danken, daß sie sich bemüht haben, hier ein im ganzen wohl gelungenes Symposium zustande zu bringen. Mein herzlicher Dank gilt Ihnen, und ehe ich Sie entlasse, möchte ich doch noch in den Symbolen der Stadt und in den Symbolen des Einflusses und Ausflusses sprechen und zwar auf Grund der gestern von Herrn Bücher kreierten biochemischen Heraldik. Die biochemische Heraldik kam bei gutem Aufpassen bei dem Vortrag von Herrn Ussing dadurch zustande, daß dieses M und o den outflux gezeigt hat. Das o muß ja unten stehen. Und wir kommen nunmehr zu dem Outflux unserer Person selbst hier aus Mosbach. Und da müßte man eigentlich auch nach dem Einflux fragen. Und dieser Einflux müßte dann eigentlich Miesbach heißen. Nun glaube ich aber nicht, daß wir den Einflux nach Miesbach verlegen können. Bei aller Verehrung, die ich für die sehr intensiven Münchener Biochemiker habe, glaube ich sogar, daß sie lieber zu uns nach Mosbach kommen. Ich meine, Mosbach ist doch eine gewisse, mehr oder weniger im geheimen blühende Zelle, die gewiß auch noch Auxin oder Kinetin haben könnte. Dieses Mosbach wird uns nun wirklich entlassen, und ich hoffe, daß wir bei dem Ausflux dann einen aktiven Transport erfahren haben in bezug auf die verschiedenen möglichen Anregungen und ein entsprechend hohes Niveau erreicht haben, wenn wir uns jetzt voneinander trennen. Ich schließe die Sitzung.